IN ASSOCIATION WITH

✕SQA

HODDER GIBSON

Model Papers

WITH ANSWERS

PLUS: Official SQA 2014 & 2015
Past Papers With Answers

National 5
Chemistry

Model Papers, 2014 & 2015 Exams

HODDER GIBSON
AN HACHETTE UK COMPANY

This book contains the official SQA 2014 and 2015 Exams for National 5 Chemistry, with associated SQA approved answers modified from the official marking instructions that accompany the paper.

In addition the book contains model papers, together with answers, plus study skills advice. These papers, some of which may include a limited number of previously published SQA questions, have been specially commissioned by Hodder Gibson, and have been written by experienced senior teachers and examiners in line with the new National 5 syllabus and assessment outlines, Spring 2013. This is not SQA material but has been devised to provide further practice for National 5 examinations in 2014 and beyond.

Hodder Gibson is grateful to the copyright holders, as credited on the final page of the Answer Section, for permission to use their material. Every effort has been made to trace the copyright holders and to obtain their permission for the use of copyright material. Hodder Gibson will be happy to receive information allowing us to rectify any error or omission in future editions.

Hachette UK's policy is to use papers that are natural, renewable and recyclable products and made from wood grown in sustainable forests. The logging and manufacturing processes are expected to conform to the environmental regulations of the country of origin.

Orders: please contact Bookpoint Ltd, 130 Park Drive, Milton Park, Abingdon, Oxon OX14 4SE. Telephone: (44) 01235 827720. Fax: (44) 01235 400454. Lines are open 9.00–5.00, Monday to Saturday, with a 24-hour message answering service. Visit our website at www.hoddereducation.co.uk. Hodder Gibson can be contacted direct on: Tel: 0141 848 1609; Fax: 0141 889 6315; email: hoddergibson@hodder.co.uk

This collection first published in 2015 by
Hodder Gibson, an imprint of Hodder Education,
An Hachette UK Company
2a Christie Street
Paisley PA1 1NB

Typeset by Aptara, Inc.

Printed in the UK

A catalogue record for this title is available from the British Library

ISBN: 978-1-4718-6052-2

3 2 1

2016 2015

Introduction

Study Skills – what you need to know to pass exams!

Pause for thought

Many students might skip quickly through a page like this. After all, we all know how to revise. Do you really though?

Think about this:

"IF YOU ALWAYS DO WHAT YOU ALWAYS DO, YOU WILL ALWAYS GET WHAT YOU HAVE ALWAYS GOT."

Do you like the grades you get? Do you want to do better? If you get full marks in your assessment, then that's great! Change nothing! This section is just to help you get that little bit better than you already are.

There are two main parts to the advice on offer here. The first part highlights fairly obvious things but which are also very important. The second part makes suggestions about revision that you might not have thought about but which WILL help you.

Part 1

DOH! It's so obvious but …

Start revising in good time

Don't leave it until the last minute – this will make you panic.

Make a revision timetable that sets out work time AND play time

Sleep and eat!

Obvious really, and very helpful. Avoid arguments or stressful things too – even games that wind you up. You need to be fit, awake and focused!

Know your place!

Make sure you know exactly **WHEN and WHERE** your exams are.

Know your enemy!

Make sure you know what to expect in the exam.

How is the paper structured?

How much time is there for each question?

What types of question are involved?

Which topics seem to come up time and time again?

Which topics are your strongest and which are your weakest?

Are all topics compulsory or are there choices?

Learn by DOING!

There is no substitute for past papers and practice papers – they are simply essential! Tackling this collection of papers and answers is exactly the right thing to be doing as your exams approach.

Part 2

People learn in different ways. Some like low light, some bright. Some like early morning, some like evening / night. Some prefer warm, some prefer cold. But everyone uses their BRAIN and the brain works when it is active. Passive learning – sitting gazing at notes – is the most INEFFICIENT way to learn anything. Below you will find tips and ideas for making your revision more effective and maybe even more enjoyable. What follows gets your brain active, and active learning works!

Activity 1 – Stop and review

Step 1

When you have done no more than 5 minutes of revision reading STOP!

Step 2

Write a heading in your own words which sums up the topic you have been revising.

Step 3

Write a summary of what you have revised in no more than two sentences. Don't fool yourself by saying, "I know it, but I cannot put it into words". That just means you don't know it well enough. If you cannot write your summary, revise that section again, knowing that you must write a summary at the end of it. Many of you will have notebooks full of blue/black ink writing. Many of the pages will not be especially attractive or memorable so try to liven them up a bit with colour as you are reviewing and rewriting. **This is a great memory aid, and memory is the most important thing.**

Activity 2 – Use technology!

Why should everything be written down? Have you thought about "mental" maps, diagrams, cartoons and colour to help you learn? And rather than write down notes, why not record your revision material?

What about having a text message revision session with friends? Keep in touch with them to find out how and what they are revising and share ideas and questions.

Why not make a video diary where you tell the camera what you are doing, what you think you have learned and what you still have to do? No one has to see or hear it, but the process of having to organise your thoughts in a formal way to explain something is a very important learning practice.

Be sure to make use of electronic files. You could begin to summarise your class notes. Your typing might be slow, but it will get faster and the typed notes will be easier to read than the scribbles in your class notes. Try to add different fonts and colours to make your work stand out. You can easily Google relevant pictures, cartoons and diagrams which you can copy and paste to make your work more attractive and **MEMORABLE**.

Activity 3 – This is it. Do this and you will know lots!

Step 1

In this task you must be very honest with yourself! Find the SQA syllabus for your subject (www.sqa.org.uk). Look at how it is broken down into main topics called MANDATORY knowledge. That means stuff you MUST know.

Step 2

BEFORE you do ANY revision on this topic, write a list of everything that you already know about the subject. It might be quite a long list but you only need to write it once. It shows you all the information that is already in your long-term memory so you know what parts you do not need to revise!

Step 3

Pick a chapter or section from your book or revision notes. Choose a fairly large section or a whole chapter to get the most out of this activity.

With a buddy, use Skype, Facetime, Twitter or any other communication you have, to play the game "If this is the answer, what is the question?". For example, if you are revising Geography and the answer you provide is "meander", your buddy would have to make up a question like "What is the word that describes a feature of a river where it flows slowly and bends often from side to side?".

Make up 10 "answers" based on the content of the chapter or section you are using. Give this to your buddy to solve while you solve theirs.

Step 4

Construct a wordsearch of at least 10 × 10 squares. You can make it as big as you like but keep it realistic. Work together with a group of friends. Many apps allow you to make wordsearch puzzles online. The words and phrases can go in any direction and phrases can be split. Your puzzle must only contain facts linked to the topic you are revising. Your task is to find 10 bits of information to hide in your puzzle, but you must not repeat information that you used in Step 3. DO NOT show where the words are. Fill up empty squares with random letters. Remember to keep a note of where your answers are hidden but do not show your friends. When you have a complete puzzle, exchange it with a friend to solve each other's puzzle.

Step 5

Now make up 10 questions (not "answers" this time) based on the same chapter used in the previous two tasks. Again, you must find NEW information that you have not yet used. Now it's getting hard to find that new information! Again, give your questions to a friend to answer.

Step 6

As you have been doing the puzzles, your brain has been actively searching for new information. Now write a NEW LIST that contains only the new information you have discovered when doing the puzzles. Your new list is the one to look at repeatedly for short bursts over the next few days. Try to remember more and more of it without looking at it. After a few days, you should be able to add words from your second list to your first list as you increase the information in your long-term memory.

FINALLY! Be inspired...

Make a list of different revision ideas and beside each one write **THINGS I HAVE** tried, **THINGS I WILL** try and **THINGS I MIGHT** try. Don't be scared of trying something new.

And remember – "FAIL TO PREPARE AND PREPARE TO FAIL!"

National 5 Chemistry

The course

Before sitting your National 5 Chemistry examination, you must have passed three **Unit Assessments** within your school or college, and produced an additional short report (approximately 100 words).

To achieve a pass in National 5 Chemistry there are then two further main components.

Component 1 – Assignment

You are required to submit an assignment that is worth 20% (20 marks) of your final grade. This assignment will be based on research and may include an experiment. This assignment requires you to apply skills, knowledge and understanding to investigate a relevant topic in chemistry and its effect on the environment and/or society. Your school or college will provide you with a Candidate's Guide for this assignment, which has been produced by the SQA. This guide gives guidance on what is required to complete the 400–800 word report and gain as many marks as possible.

Your assignment report will be marked by the SQA.

Component 2 – The Question Paper

The question paper will assess breadth and depth of knowledge and understanding from across all of the three Units. The question paper will require you to:

- Make statements, provide explanations, and describe information to demonstrate knowledge and understanding.
- Apply knowledge and understanding to new situations to solve problems.
- Plan and design experiments.
- Present information in various forms such as graphs, tables etc.
- Perform calculations based on information given.
- Give predictions or make generalisations based on information given.
- Draw conclusions based on information given.
- Suggest improvement to experiments to improve the accuracy of results obtained or to improve safety.

To achieve a "C" grade in National 5 Chemistry you must achieve about 50% of the 100 marks available when the two components, i.e. the Question Paper and the Assignment are combined. For a B you will need 60%, while for an "A" grade you must ensure that you gain as many of the available marks as possible, and at least 70%.

This book contains model papers that cover the content of the National 5 Chemistry course and illustrate the standard, structure and requirements of the Question Paper that you will sit during your exam.

Each model paper consists of two sections. (A marking scheme for each section is provided at the end of this book.)

- Section A will contain objective questions (multiple choice) and will have 20 marks.
- Section B will contain restricted and extended response questions and will have 60 marks.

Each model paper contains a variety of questions including some that require:

- demonstration and application of knowledge, and understanding of the mandatory content of the course from across the three units
- application of scientific inquiry skills.

How to use this book

This book can be used in two ways:

1. You can complete an entire model paper under exam conditions, without the use of books or notes, and then mark the papers using the marking scheme provided. This method gives you a clear indication of the level you are working at and should highlight the content areas that you need to work on before attempting the next model paper. This method also allows you to see your progress as you complete each model paper.

2. You can complete a model paper using your notes and books. Try the question first and then refer to your notes if you are unable to answer the question. This is a form of studying and by doing this you will cover all the areas of content that you are weakest in. You should notice that you are referring to your notes less with each model paper completed.

Try to practise as many questions as possible. This will get you used to the language used in the Question Papers and ultimately improve your chances of success.

Some hints and tips

Below is a list of hints and tips that will help you to achieve your full potential in the National 5 exam.

- Ensure that you **read each question carefully**. Scanning the question and missing the main points results in mistakes being made. Some students highlight the main points of a question with a highlighter pen to ensure that they don't miss anything out.

- Open ended questions include the statement **"Using your knowledge of chemistry"**. These questions provide you with an opportunity to show off your chemistry knowledge. To obtain the three marks on offer for these questions, you must demonstrate a good understanding of the chemistry involved and provide a logically correct answer to the question posed.

- When doing calculations, ensure that you **show all of your working**. If you make a simple arithmetical mistake you may still be awarded some of the marks, but only if your working is laid out clearly so that the examiner can see where you went wrong and what you did correctly. Just giving the answers is very risky so you should always show your working.

- **Attempt all questions.** Giving no answer at all means that you will definitely not gain any marks.

- When you are required to read a passage to answer a question, ensure that you **read it carefully** as the information you require is contained within it. It may not be obvious at first, but the answers will be contained within the passage.

- If you are asked to "explain" in a question, then you must **explain your answer fully**. For example, if you are asked to explain how a covalent bond holds atoms together then you cannot simply say:

    "A covalent bond is a shared pair of electrons between atoms in a non-metal."

This answer tells the examiner what a covalent bond is, but does not explain how it holds the atoms together. To gain the marks, an answer similar to this should be written:

    "A covalent bond is a shared pair of electrons between atoms in a non-metal. The shared electrons are attracted to the nuclei of both atoms, which creates a tug-of-war effect creating the covalent bond."

- You will be required to draw one graph in each exam. To obtain all the marks, ensure that the graphs have **labels, units, points plotted correctly** and a line of "best fit" drawn between the points.

- Use your **data booklet** when you are asked to write formulas, ionic formulas, formula mass etc. You have the data booklet in front of you so use it to double check the numbers you require.

- Work on your **timing**. The multiple-choice section (Section 1) should take approximately 30 minutes. Attempt to answer the multiple-choice questions before you look at the four possible answers, as this will improve your confidence. Use scrap paper when required to scribble down structural formulae, calculations, chemical formulae etc., as this will reduce your chance of making errors. If you are finding the question difficult, try to eliminate the obviously wrong answers to increase your chances.

- When asked to **predict or estimate** based on information from a graph or a table, then take your time to look for patterns. For example, if asked to predict a boiling point, try to establish if there is a regular change in boiling point and use that regular pattern to establish the unknown boiling point.

- When drawing a **diagram** of an experiment ask yourself the question, "Would this work if I set it up exactly like this in the lab?" Ensure that the method you have drawn would produce the desired results safely. If, for example, you are heating a flammable reactant such as alcohol then you will not gain the marks if you heat it with a Bunsen burner in your diagram; a water bath would be much safer! Make sure your diagram is labelled clearly.

Good luck!

Remember that the rewards for passing National 5 Chemistry are well worth it! Your pass will help you get the future you want for yourself. In the exam, be confident in your own ability. If you're not sure how to answer a question, trust your instincts and just give it a go anyway. Keep calm and don't panic! GOOD LUCK!

Model Paper 1

Whilst this Model Paper has been specially commissioned by Hodder Gibson for use as practice for the National 5 exams, the key reference documents remain the SQA Specimen Paper 2013 and the SQA Past Papers 2014 and 2015.

HODDER GIBSON
LEARN MORE

National Qualifications
MODEL PAPER 1

Chemistry
Section 1—Questions

Duration — 2 hours

Instructions for completion of Section 1 are given on Page two of the question paper.

Record your answers on the grid on Page three of your answer booklet.

Do not write in this booklet.

Before leaving the examination room you must give your answer booklet to the Invigilator. If you do not, you may lose all the marks for this paper.

SECTION 1

1. Which of the following elements has a covalent molecular structure?

    A   Sodium

    B   Helium

    C   Silicon

    D   Hydrogen

2. An element has an atomic number of 11 and a mass number of 23. The number of electrons present in an atom of this element is?

    A   11

    B   12

    C   23

    D   34

3. Ionic compounds conduct in solution because

    A   the ions are free to move

    B   the ions are not free to move

    C   the electrons are free to move

    D   the electrons are not free to move

4. Which of the following compounds has molecules with the same shape as ammonia?

    A   Carbon dioxide

    B   Hydrogen oxide

    C   Sulfur dioxide

    D   Phosphorus hydride

5. Which line in the table correctly describes a neutron?

	Mass	Charge
A	1	−1
B	negligible	0
C	1	+1
D	1	0

Questions **6** and **7** refer to the table below.

The table contains information about some substances.

Substance	Melting point ( ) ^0C	Boiling point ( ) ^0C	Conducts as a solid	Conducts as a liquid
A	-7	59	No	No
B	1492	2897	Yes	Yes
C	1407	2357	No	No
D	606	1305	No	Yes

6.   Identify the substance that is a liquid at room temperature (21 °C).

7.   Identify the substance that exists as a covalent network.

8.   Which of the following statements correctly describes the concentration of $H^+(aq)$ and OH (aq) ions in an acidic substance when compared to pure water?

   A   The concentration of $H'(aq)$ and $OH^-(aq)$ ions are zero.

   B   The concentration of $H^+(aq)$ and $OH^-(aq)$ ions are equal.

   C   The concentration of $H^+(aq)$ is higher than $OH^-(aq)$ ions.

   D   The concentration of $H^+(aq)$ is lower than $OH^-(aq)$ ions.

9.

$$\begin{array}{ccccccccc}
 & H & & H & & & & H & \\
 & | & & | & & & & | & \\
H & C & - & C & - & C & - & C & - & C & - H \\
 & | & & | & & | & & | & & | \\
 & H & & H & & H & & H & & H
\end{array}$$

The name of the above compound is

   A   but-2-ene

   B   pent-2-ene

   C   but-3-ene

   D   pent-3-ene.

10. Shown below is the structure of a compound known as neopentane.

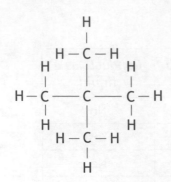

The systematic name of the above compound is

A   2,3-dimethylbutane

B   3,2-dimethylbutane

C   2,2-dimethylpropane

D   3,2-dimethylpropane.

11. Which of the following structures belongs to the same homologous series as the compound with the formula C_3H_8?

A

```
    H   H
    |   |
H — C — C — H
    |   |
H — C — C — H
    |   |
    H   H
```

B

```
    H               H
    |               |
H — C — C = C — C — H
    |   |   |   |
    H   H   H   H
```

C

```
            H
            |
        H — C — H
    H   H   |   H
    |   |   |   |
H — C — C — C — C — H
    |   |   |   |
    H   H   H   H
```

D

```
            H
            |
        H — C — H
    H   H   |
    |   |   |
H — C — C — C = C — H
    |   |       |
    H   H       H
```

12. Which of the following compounds is **not** an isomer of pent-1-ene?

A   but-1-ene

B   pent-2-ene

C   cyclopentane

D   2-methylbut-1-ene

13. Identify the hydrocarbon that reacts quickly with bromine solution.

A

B

C

D

14. Metallic bonds are due to

   A   a shared pair of electrons

   B   an attraction between positive ions and negative ions

   C   an attraction between positive ions and delocalised electrons

   D   an attraction between negative ions and delocalised electrons.

15. $Cu^{2+} + 2e^- \rightarrow Cu$

   This ion electron equation represents the

   A   reduction of copper(II) ions

   B   reduction of copper(I) ions

   C   oxidation of copper(II) ions

   D   oxidation of copper(I) ions.

16. Which of the following metals can be obtained from its ore by heat alone?

   A   Iron

   B   Potassium

   C   Mercury

   D   Aluminium

Page five

17.

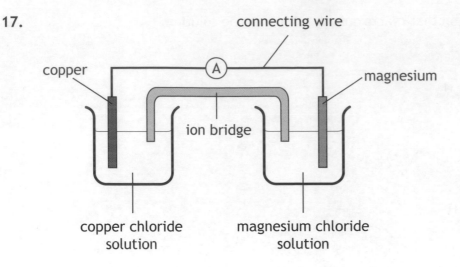

In the cell shown, electrons flow through

A   the solution from copper to magnesium

B   the solution from magnesium to copper

C   the connecting wire from copper to magnesium

D   the connecting wire from magnesium to copper.

18.   Four cells were made by joining magnesium, aluminium, nickel and zinc to lead. The voltages are shown in the table.

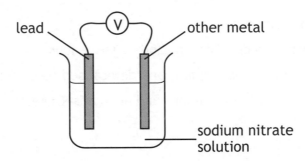

Which line in the table below shows the voltage of the cell containing magnesium joined to lead?

You may wish to use the data booklet to help you.

Cell	Voltage (V)
A	0·3
B	1·0
C	1·1
D	1·2

19. The half-life of the isotope ^{210}Pb is 22 years.

What fraction of the original sample will remain after 44 years?

A $\frac{1}{2}$

B $\frac{1}{4}$

C $\frac{1}{8}$

D $\frac{1}{16}$

20. The structure shows a section of the addition polymer Teflon.

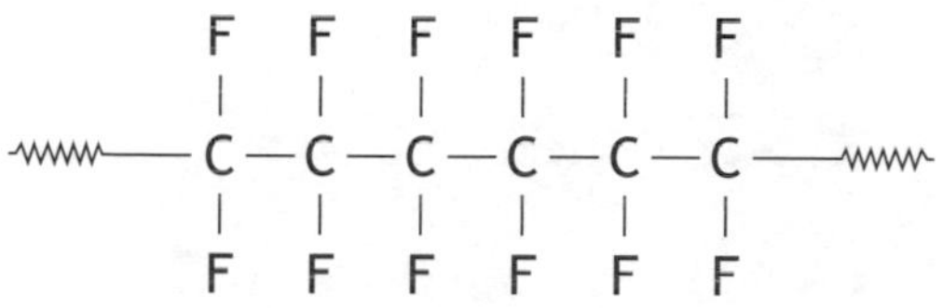

Which molecule is used to make this polymer?

A

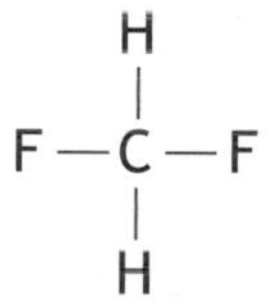

B

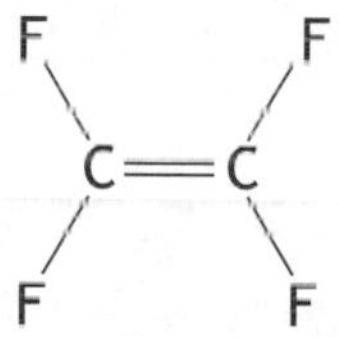

C

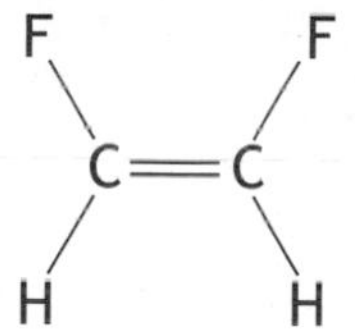

D

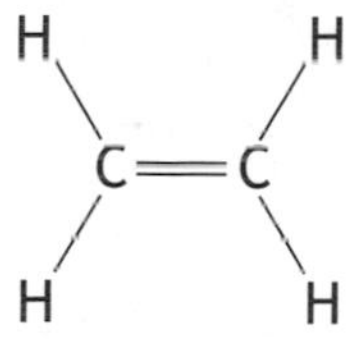

**[END OF SECTION 1. NOW ATTEMPT THE QUESTIONS IN SECTION 2
OF YOUR QUESTION AND ANSWER BOOKLET.]**

National Qualifications MODEL PAPER 1

Chemistry
Section 1—Answer
Grid and Section 2

Duration — 2 hours

Total marks — 80

SECTION 1 — 20 marks

Attempt ALL questions in this section.

Instructions for completion of Section 1 are given on Page two.

SECTION 2 — 60 marks

Attempt ALL questions in this section.

Read all questions carefully before attempting.

Use **blue** or **black** ink. Do NOT use gel pens.

Write your answers in the spaces provided. Additional space for answers and rough work is provided at the end of this booklet. If you use this space, write clearly the number of the question you are attempting. Any rough work must be written in this booklet. You should score through your rough work when you have written your fair copy.

HODDER GIBSON
LEARN MORE

SECTION 1— 20 marks

The questions for Section 1 are contained in the booklet Chemistry Section 1—Questions.
Read these and record your answers on the grid on Page three opposite.

1. The answer to each question is **either** A, B, C or D. Decide what your answer is, then fill in the appropriate bubble (see sample question below).

2. There is **only one correct** answer to each question.

3. Any rough working should be done on the additional space for rough working and answers sheet.

Sample Question

To show that the ink in a ball-pen consists of a mixture of dyes, the method of separation would be:

    A   fractional distillation

    B   chromatography

    C   fractional crystallisation

    D   filtration.

The correct answer is **B**—chromatography.   The answer **B** bubble has been clearly filled in (see below).

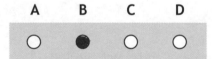

Changing an answer

If you decide to change your answer, cancel your first answer by putting a cross through it (see below) and fill in the answer you want. The answer below has been changed to **D**.

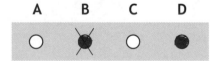

If you then decide to change back to an answer you have already scored out, put a tick (✓) to the **right** of the answer you want, as shown below:

A   B   C   D   A   B   C   D

○   ✗✓   ○   ✗   or   ○   ✗✓   ○   ○

SECTION 1—Answer Grid

	A	B	C	D
1	○	○	○	○
2	○	○	○	○
3	○	○	○	○
4	○	○	○	○
5	○	○	○	○
6	○	○	○	○
7	○	○	○	○
8	○	○	○	○
9	○	○	○	○
10	○	○	○	○
11	○	○	○	○
12	○	○	○	○
13	○	○	○	○
14	○	○	○	○
15	○	○	○	○
16	○	○	○	○
17	○	○	○	○
18	○	○	○	○
19	○	○	○	○
20	○	○	○	○

[BLANK PAGE]

MARKS | DO NOT WRITE IN THIS MARGIN

SECTION 2— 60 marks

Attempt ALL questions.

1. Rapid inflation of airbags in cars is caused by the production of nitrogen gas.

    The graph gives information on the volume of gas produced over 30 microseconds.

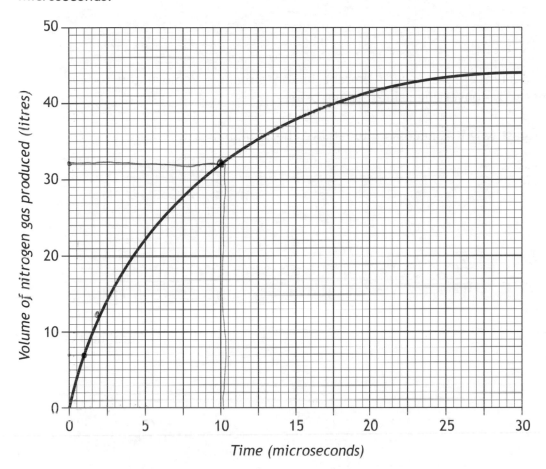

    (a)   (i)   Calculate the average rate of reaction, in litres per microsecond, between 2 and 10 microseconds.   **2**

          (ii) At what time, in microseconds, has half of the final volume of nitrogen gas been produced?   **1**

MARKS | DO NOT WRITE IN THIS MARGIN

1. (continued)

(b) In some types of airbag, electrical energy causes sodium azide, NaN_3, to decompose producing sodium metal and nitrogen gas.

Using symbols and formulae, write an equation for this reaction.   1

There is no need to balance this equation.

(c) Potassium nitrate is also present in the airbag to remove the sodium metal by converting it into sodium oxide.

Suggest why it is necessary to remove the sodium metal.   1

Total marks   5

MARKS

DO NOT WRITE IN THIS MARGIN

2. Fuels have developed greatly in the past 200 years. More traditional fuels such as candle wax, peat and coal were the most commonly used fuels in the 19th century.

Using your knowledge of chemistry, describe how you could establish which of these fuels was the most efficient and which produced the least pollution.

3

Total marks   3

MARKS | DO NOT WRITE IN THIS MARGIN

3. Egg shells are made up mainly of calcium carbonate. A pupil carried out an experiment to react egg shell with dilute hydrochloric acid. A gas was produced.

    (a) Complete the diagram to show the apparatus which could have been used to collect and measure the volume of gas produced.   1

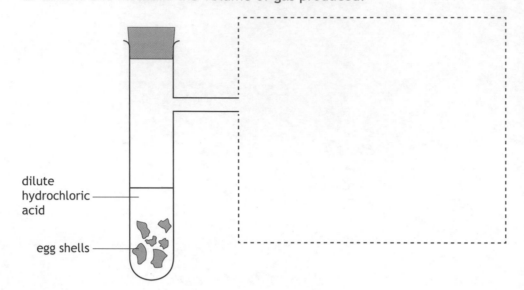

    (b) Name the salt produced in this reaction.   1

    (c) The volume of gas produced during the reaction was measured.

Time (min)	Volume of gas (cm³)
0	0
2	47
4	92
6	114
8	118
10	118

MARKS DO NOT WRITE IN THIS MARGIN

3. (c) (continued)

Draw a line graph of volume of gas against time. Use appropriate scales to fill most of the paper. **3**

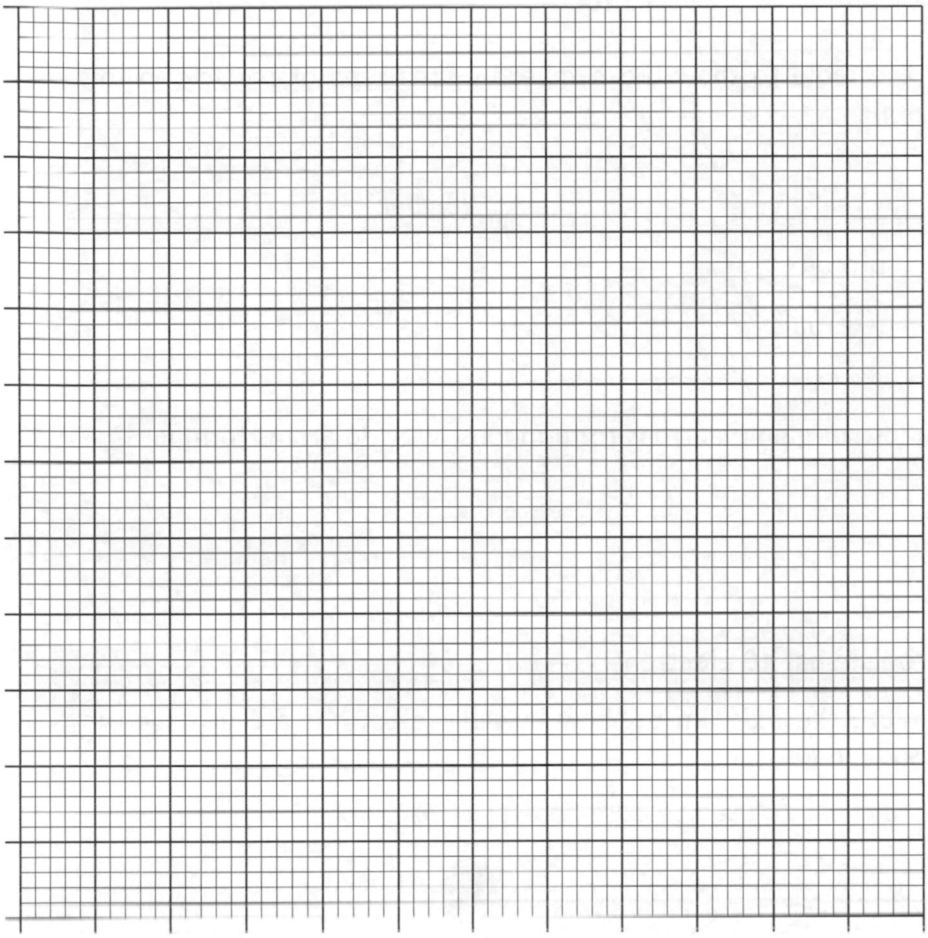

Total marks 5

MARKS | DO NOT WRITE IN THIS MARGIN

4. Atoms contain particles called protons, neutrons and electrons.

The nuclide notation of a sodium atom is shown.

$$^{24}_{11}Na$$

(a) Complete the table to show the number of each type of particle in this sodium atom.

1

Particle	Number
proton	
neutron	

(b) Atoms can lose or gain electrons to form ions. Why do atoms form ions?

1

(c) An ion of sodium has 10 electrons.

(i) Complete the diagram to show how the electrons are arranged in this sodium ion.

1

You may wish to use the data booklet to help you.

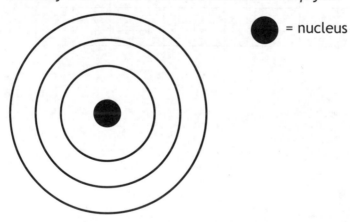

= nucleus

(ii) Explain what holds the negatively charged electrons in place around the nucleus.

1

Total marks 4

MARKS   DO NOT WRITE IN THIS MARGIN

5. The octane number of petrol is a measure of how efficiently it burns as a fuel.

The higher the octane number, the more efficient the fuel.

(a) The octane numbers for some hydrocarbons are shown.

Hydrocarbon	Number of carbon atoms	Octane number
butane	4	90
pentane	5	62
hexane	6	
heptane	7	0
octane	8	−19
2-methylpentane	6	71
2-methylhexane	7	44
2-methylheptane	8	23

    (i) Predict the octane number for hexane.   **1**

    (ii) State a relationship between the structure of the hydrocarbon and their efficiency.   **1**

(b) A student investigated the amount of energy released when hexane was burned. The student recorded the following data.

Mass of hexane burned	5 g
Volume of water	1 litre
Initial temperature of water	20 °C
Final temperature of water	78 °C
Specific heat capacity of water	4.18 kJ kg °C^{-1}

Calculate the energy released, in kJ.   **3**

You may wish to use the data booklet to help you.

Total marks   **5**

MARKS | DO NOT WRITE IN THIS MARGIN

6. Read the passage below and answer the questions that follow.

Ocean Dead Zones

A dead zone is an area of an ocean (or lake) that has too little oxygen to support marine life; it is hypoxic. This is a natural phenomenon that has been increasing in shallow coastal and estuarine areas as a result of human activities.

Dead zones form due to an increase in nutrients in the water (particularly phosphorus and nitrogen). Human activities have resulted in the near doubling of nitrogen and tripling of phosphorus flows to the environment when compared to natural values.

This dramatic increase in previously limited nutrients results in massive algal blooms. These "red tides" or harmful Algal Blooms can kill fish, cause human illness through shellfish poisoning, and death of marine mammals and shore birds.

This population explosion of algae is unsustainable, and eventually the algae die off, as they block out the light and use up all the oxygen. The algae sink to the bottom, and bacterial decomposition uses the remaining oxygen from the water.

The passage on Ocean Dead Zones was taken from an article published on "sailorsforthesea.org".

(a) Suggest which human activities result in nitrogen and phosphorus compounds found in the water.   1

(b) Name the term used to describe an area of the ocean that does not have enough oxygen to support life.   1

(c) Name the two factors that directly contribute to the low oxygen levels found in these dead zones.   1

(d) How does the concentration of phosphorus in 'Dead Zones' compare to natural levels of phosphorus found in water?   1

Total marks   4

MARKS | DO NOT WRITE IN THIS MARGIN

7. The voltage obtained when different pairs of metal strips are connected in a cell varies and this leads to the electrochemical series.

Using the apparatus below, a student investigated the electrochemical series. Copper and four other metal strips were used in this investigation.

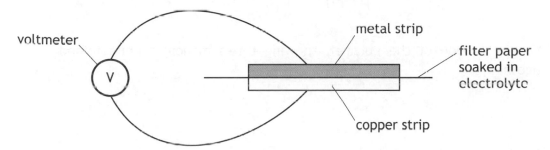

The results are shown.

Metal strip	Voltage (V)	Direction of electron flow
1	0·6	metal 1 to copper
2	0·2	copper to metal 2
3	0·9	metal 3 to copper
4	0·1	copper to metal 4

(a) Circle the correct metal to complete the sentence.   1

Connecting metal   1   to metal 2 would produce the largest voltage.
                          3
                          4

(b) What would be the reading on the voltmeter if both strips of metal were copper?   1

(c) Why can glucose ($C_6H_{12}O_6$) solution **not** be used as the electrolyte?   1

Total marks   3

MARKS | DO NOT WRITE IN THIS MARGIN

8. The monomer in superglue has the following structure.

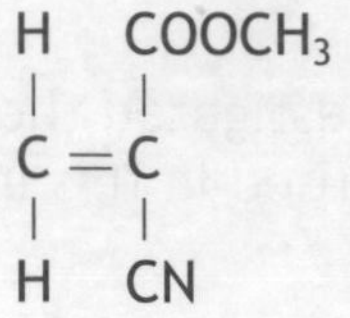

$$
\begin{array}{c}
\text{H} \quad \text{COOCH}_3 \\
| \qquad | \\
\text{C} = \text{C} \\
| \qquad | \\
\text{H} \quad \text{CN}
\end{array}
$$

(a) Draw a section of the polymer, showing **three** monomer units joined together.

1

(b) Name the type of polymerisation that takes place when these monomers combine.

1

(c) Bromine reacts with the monomer to produce a saturated compound.

Draw a structural formula for this compound.

1

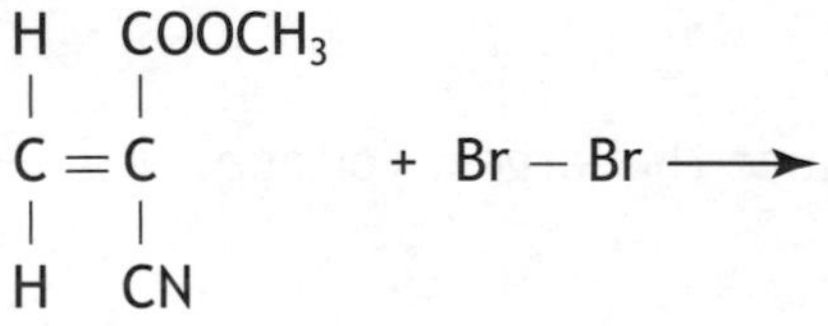

$$
\begin{array}{c}
\text{H} \quad \text{COOCH}_3 \\
| \qquad | \\
\text{C} = \text{C} \qquad\qquad + \; \text{Br} - \text{Br} \longrightarrow \\
| \qquad | \\
\text{H} \quad \text{CN}
\end{array}
$$

Total marks   3

MARKS | DO NOT WRITE IN THIS MARGIN

9. Chlorofluorocarbons (CFCs) are a family of compounds that are highly effective as refrigerants and aerosol propellants. However, they are known to damage the ozone layer, and can have a long atmospheric life, which is a measure of how long they exist in the atmosphere.

One example of a CFC molecule (CCl_2F_2) is shown.

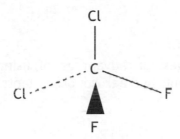

(a) What term is used to describe the **shape** of this molecule?   1

(b) Scientists have developed compounds to replace CFCs. The table shows information about the ratio of atoms in CCl_2F_2 and compounds used to replace it.

Compound	Number of atoms				Atmospheric life (years)
	C	Cl	F	H	
CCl_2F_2	1	2	2	0	102
Replacement 1	1	1	2	1	13.3
Replacement 2	2	0	4	2	14.6
Replacement 3	1	0	2	3	5.6

(i) Draw a structural formula for Replacement 2.   1

MARKS | DO NOT WRITE IN THIS MARGIN

9. (b) (continued)

(ii) Compared with CCl_2F_2, the replacement compounds contain less of which element? **1**

(iii) From the table, suggest an advantage of using the replacement molecules as refrigerants and aerosol propellants. **1**

Total marks 4

MARKS | DO NOT WRITE IN THIS MARGIN

10. Silver jewellery slowly tarnishes in air. This is due to the formation of silver(I) sulfide.

The silver(I) sulfide can be converted back to silver using the following apparatus.

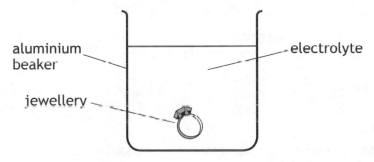

The equation for the reaction that takes place in the beaker is shown.

$$3Ag_2S(aq) + 2Al(s) \longrightarrow 6Ag(s) + Al_2S_3(aq)$$

(a) Calculate the mass, in grams, of silver produced when 0.135 g of aluminium is used up.   3

(b) How would you show that aluminium has been lost from the beaker during this reaction?   1

Total marks   4

MARKS | DO NOT WRITE IN THIS MARGIN

11. Ethanol is a member of the alcohol family of compounds.

(a) Ethanol can be manufactured from ethene as shown in the following hydration reaction.

$$\underset{\underset{H}{\overset{H}{|}}}{\overset{\overset{H}{|}}{C}} = \underset{\underset{H}{\overset{H}{|}}}{\overset{\overset{H}{|}}{C}} + H_2O \xrightarrow{\text{catalyst}} H - \underset{\underset{H}{\overset{H}{|}}}{\overset{\overset{H}{|}}{C}} - \underset{\underset{OH}{\overset{H}{|}}}{\overset{\overset{H}{|}}{C}} - H$$

What other name can be given to this type of reaction?   1

(b) The following compound is used to give ice cream a rum flavour. It is produced by reacting ethanol with propanoic acid.

$$H - \underset{\underset{H}{\overset{H}{|}}}{\overset{\overset{H}{|}}{C}} - \underset{\underset{H}{\overset{H}{|}}}{\overset{\overset{H}{|}}{C}} - O - \overset{\overset{O}{||}}{C} - \underset{\underset{H}{\overset{H}{|}}}{\overset{\overset{H}{|}}{C}} - \underset{\underset{H}{\overset{H}{|}}}{\overset{\overset{H}{|}}{C}} - H$$

(i) Draw the full structural formula of propanoic acid.   1

(ii) To which homologous series does propanoic acid belong?   1

(c) Butan-2-ol is another member of the alkanol family.

$$H - \underset{\underset{H}{\overset{H}{|}}}{\overset{\overset{H}{|}}{C}} - \underset{\underset{OH}{\overset{H}{|}}}{\overset{\overset{H}{|}}{C}} - \underset{\underset{H}{\overset{H}{|}}}{\overset{\overset{H}{|}}{C}} - \underset{\underset{H}{\overset{H}{|}}}{\overset{\overset{H}{|}}{C}} - H$$

Draw the full structural formula for an isomer of butan-2-ol.   1

Total marks   4

MARKS | DO NOT WRITE IN THIS MARGIN

12. Uranium is a silvery white metallic element that is radioactive because all its isotopes are unstable. Uranium decays by alpha emission.

(a) What is meant by the term isotope?   **1**

(b) Write a balanced nuclear equation for the alpha decay of $^{238}_{92}U$ **2**

(c) Uranium-238 has a half-life of 4.5 billion years.

What is meant by the term half-life?   **1**

Total Marks   4

13. Read the following passage carefully then answer the question that follows.

Hydrogen doesn't fit

The first element, hydrogen, has been causing trouble for some time. It can be placed in group 1, as it usually is, or with the halogens in group 7.

Periodic Table of the Elements

																	2 Helium He
3 Lithium Li	4 Beryllium Be		Atomic number Name of element Symbol				1 Hydrogen H					5 Boron B	6 Carbon C	7 Nitrogen N	8 Oxygen O	9 Fluorine F	10 Neon Ne
11 Sodium Na	12 Magnesium Mg											13 Aluminum Al	14 Silicon Si	15 Phosphorus P	16 Sulphur S	17 Chlorine Cl	18 Argon Ar
19 Potassium K	20 Calcium Ca	21 Scandium Sc	22 Titanium Ti	23 Vanadium V	24 Chromium Cr	25 Manganese Mn	26 Iron Fe	27 Cobalt Co	28 Nickel Ni	29 Copper Cu	30 Zinc Zn	31 Gallium Ga	32 Germanium Ge	33 Arsenic As	34 Selenium Se	35 Bromine Br	36 Krypton Kr
37 Rubidium RB	38 Strontium Sr	39 Yttrium Y	40 Zirconium Zr	41 Niobium Nb	42 Molybdenum Mu	43 Technetium Tc	44 Ruthenium Ru	45 Rhodium Rh	46 Palladium Pd	47 Silver Ag	48 Cadmium Cd	49 Indium In	50 Tin Sn	51 Antimony Sb	52 Tellurium Te	53 Iodine I	54 Xenon Xe
55 Caesium Cs	56 Barium Ba	57 Lanthanum La	58–71 ●	72 Hafnium Hf	73 Tantalum Ta	74 Tungsten W	75 Rhenium Re	76 Osmium Os	77 Iridium Ir	78 Platinum Pt	79 Gold Au	80 Mercury Hg	81 Thallium Tl	82 Lead Pb	83 Bismuth Bi	84 Polonium Po	85 Astatine At	86 Radon Rn
87 Francium Fr	88 Radium Ra	89 Actinium Ac	90–103 ■	104 Rutherfordium Rf	105 Dubnium Db	106 Seaborgium Sg	107 Bohrium Bh	108 Hassium Hs	109 Meitnerium Mt								

●	58 Cerium Ce	59 Praseodymium Pr	60 Neodymium Nd	61 Promethium Pm	62 Samarium Sm	63 Europium Eu	64 Gadolinium Gd	65 Terbium Tb	66 Dysprosium Dy	67 Holmium Ho	68 Erbium Er	69 Thulium Tm	70 Ytterbium Yb	71 Lutetium Lu
■	90 Thorium Th	91 Protactinium Pa	92 Uranium U	93 Neptunium Np	94 Plutonium Pu	95 Americium Am	96 Curium Cm	97 Berkelium Bk	98 Californium Cf	99 Einsteinium Es	100 Fermium Fm	101 Mendelevium Md	102 Nobelium No	103 Lawrencium Lr

Some authors avoid the hydrogen problem altogether by removing it from the main body and by allowing it to float above the rest of the table.

Passage from RSC.org

Using your knowledge of chemistry, give reasons why hydrogen can be placed above group 1 or group 7.

3

Total Marks   3

MARKS | DO NOT WRITE IN THIS MARGIN

14. A student carried out a titration using the chemicals and apparatus shown.

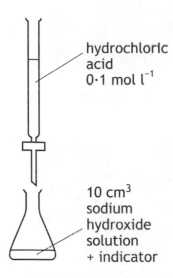

hydrochloric acid
0·1 mol l^{-1}

10 cm^3 sodium hydroxide solution + indicator

	Rough titre	1st titre	2nd titre
Initial burette reading/cm^3	0·3	0·2	0·5
Final burette reading/cm^3	26·6	25·3	25·4
Volume used/cm^3	26·3	25·1	24·9

(a) Using the results in the table, calculate the **average** volume, in cm^3, of hydrochloric acid required to neutralise the sodium hydroxide solution.   **1**

(b) The equation for the reaction is:

HCl + NaOH → NaCl + H$_2$O

Using the answer from part (a), calculate the concentration, in mol l^{-1} of the sodium hydroxide solution.   **3**

Show your working clearly.

Total Marks   4

MARKS | DO NOT WRITE IN THIS MARGIN

15. Potassium hydroxide reacts with sulfuric acid to form potassium sulfate, which can be used as a fertiliser.

$$KOH(aq) + H_2SO_4(aq) \rightarrow K_2SO_4(aq) + H_2O(l)$$

(a) Balance the above equation.　　　　1

(b) Name the type of chemical reaction taking place.　　　　1

(c) Calculate the percentage, by mass, of potassium in potassium sulfate.　　　　3

　　Show your working clearly.

Total Marks　5

[END OF MODEL PAPER]

MARKS

DO NOT WRITE IN THIS MARGIN

ADDITIONAL SPACE FOR ROUGH WORKING AND ANSWERS

MARKS

ADDITIONAL SPACE FOR ROUGH WORKING AND ANSWERS

ADDITIONAL SPACE FOR ANSWERS

Additional graph paper for Question 3 (c)

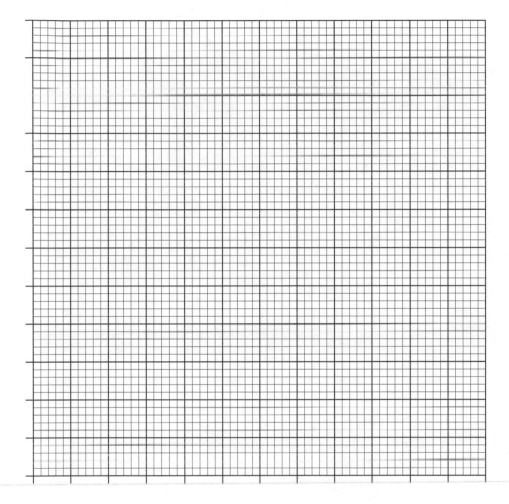

Model Paper 2

Whilst this Model Paper has been specially commissioned by Hodder Gibson for use as practice for the National 5 exams, the key reference documents remain the SQA Specimen Paper 2013 and the SQA Past Papers 2014 and 2015.

HODDER
GIBSON
LEARN MORE

National Qualifications
MODEL PAPER 2

Chemistry
Section 1—Questions

Duration — 2 hours

Instructions for completion of Section 1 are given on Page two of the question paper.

Record your answers on the grid on Page three of your answer booklet.

Do not write in this booklet.

Before leaving the examination room you must give your answer booklet to the Invigilator. If you do not, you may lose all the marks for this paper.

SECTION 1

1. Which of the following elements was discovered before 1775?

   A   Silicon

   B   Oxygen

   C   Bromine

   D   Magnesium

2. Which of the following is the electron arrangement of sodium metal?

   A   2, 8, 1

   B   2, 8, 2

   C   2, 8, 7

   D   2, 8, 8

3. Which of the following has a covalent network structure?

   A   Neon

   B   Silicon dioxide

   C   Calcium chloride

   D   Carbon dioxide

4. The shape of a methane molecule is shown.

   Which of the following compounds would have molecules the same shape as a methane molecule?

   A   Water

   B   Ammonia

   C   Sulfur dioxide

   D   Carbon tetrachloride

5. What is the charge on the copper ion in CuO?

    A   1+

    B   2+

    C   1−

    D   2−

6. Solid ionic compounds do not conduct electricity because

    A   the ions are not free to move

    B   the electrons are not free to move

    C   solid substances never conduct electricity

    D   there are no charged particles in ionic compounds.

7. Which of the following oxides dissolves in water to produce an acidic solution?

    A SO_2

    B SiO_2

    C SnO_2

    D PbO_2

8. Which of the following could be the molecular formula of a cycloalkane?

    A C_7H_{10}

    B C_7H_{12}

    C C_7H_{14}

    D C_7H_{16}

9.

The name of the above compound is

    A   2,2-dimethylbutane

    B   2-ethylpropane

    C   2-methylbutane

    D   3-methylbutane.

Page three

10.   Three members of the cycloalkene homologous series are:

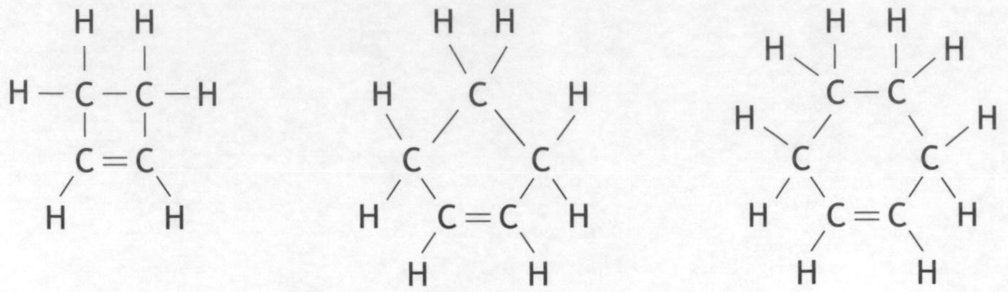

The general formula for this homologous series is

A C_nH_{2n+2}

B C_nH_{2n}

C C_nH_{2n-2}

D C_nH_{2n-4}.

11.   Which two families of compounds react together to produce esters?

A   Carboxylic acids and alcohols

B   Alkenes and alcohols

C   Alkenes and cycloalkenes

D   Carboxylic acids and cycloalkenes

12.   What is the correct systematic name for the following compound?

$CH_3CH(CH_3)CH_2CH_2CH_3$

A   Hexane

B   Pentane

C   2-methylpentane

D   3-methylpentane

13. Which of the following is an isomer of heptane?

A

```
    H   H   H   H   H
    |   |   |   |   |
H — C — C — C — C — C — H
    |   |   |       |
    H   H   H       H
                |
            H — C — H
                |
                H
```

B

```
    H   H   H   H   H   H
    |   |   |   |   |   |
H — C — C — C — C — C — C — H
    |   |   |       |   |
    H   H   H       H   H
                |
            H — C — H
                |
                H
```

C

```
                    H
                    |
                H — C — H
    H   H   H   H   |   H
    |   |   |   |       |
H — C — C — C — C — C = C — H
    |   |   |   |
    H   H   H   H
```

D

```
    H   H   H   H   H   H   H
    |   |   |   |   |   |   |
H — C — C — C — C — C = C — C — H
    |   |   |   |           |
    H   H   H   H           H
```

14. When propene undergoes an addition reaction with hydrogen bromide, two products are formed.

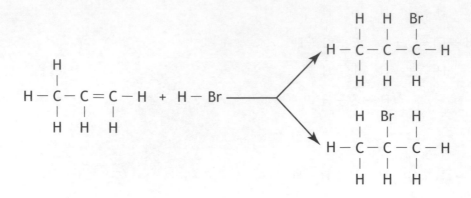

Which of the following alkenes will also produce two products when it undergoes an addition reaction with hydrogen bromide?

A   Ethene

B   But-1-ene

C   But-2-ene

D   Hex-3-ene

15. Experiments were performed on three unknown metal elements, X, Y and Z to establish their reactivity. The results of the experiments are recorded in the table below.

Metal	Reaction with water	Reaction with dilute acid
X	no reaction	no reaction
Y	slow reaction	fast reaction
Z	no reaction	slow reaction

The order of reactivity of the metals, starting with the most reactive, is

A   Y, X, Z

B   X, Z, Y

C   Y, Z, X

D   X, Y, Z.

16. Which of the following diagrams could be used to represent the structure of magnesium?

A

B

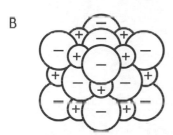

C

D

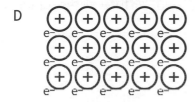

17. Which pair of metals, when connected in a cell, would give the highest voltage and a flow of electrons from **X** to **Y**?

You may wish to use the data booklet to help you.

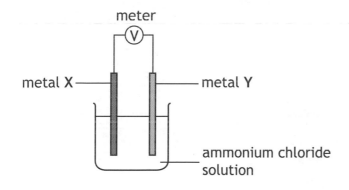

	Metal X	Metal Y
A	magnesium	copper
B	copper	magnesium
C	zinc	tin
D	tin	zinc

18. Which of the following metals would react with zinc chloride solution?

You may wish to use the data booklet to help you.

A   Copper

B   Gold

C   Iron

D   Magnesium

19. A radioisotope of thorium forms protactinium-231 by beta emission.

What is the mass number of the radioisotope of thorium?

A   230

B   231

C   232

D   235

20. Part of the structure of an addition polymer showing two different monomer units combined is shown.

$$-\overset{\displaystyle H}{\underset{\displaystyle H}{\overset{|}{\underset{|}{C}}}}-\overset{\displaystyle H}{\underset{\displaystyle H}{\overset{|}{\underset{|}{C}}}}-\overset{\displaystyle CH_3}{\underset{\displaystyle H}{\overset{|}{\underset{|}{C}}}}-\overset{\displaystyle H}{\underset{\displaystyle H}{\overset{|}{\underset{|}{C}}}}-\overset{\displaystyle H}{\underset{\displaystyle H}{\overset{|}{\underset{|}{C}}}}-\overset{\displaystyle H}{\underset{\displaystyle H}{\overset{|}{\underset{|}{C}}}}-$$

Which pair of alkenes could be used as monomers for this polymer?

A   Ethene and propene

B   Ethene and butene

C   Propene and butene

D   Ethene and pentene

**[END OF SECTION 1. NOW ATTEMPT THE QUESTIONS IN SECTION 2
OF YOUR QUESTION AND ANSWER BOOKLET.]**

N5

National Qualifications
MODEL PAPER 2

Chemistry
Section 1—Answer
Grid and Section 2

Duration — 2 hours

Total marks — 80

SECTION 1 — 20 marks

Attempt ALL questions in this section.

Instructions for completion of Section 1 are given on Page two.

SECTION 2 — 60 marks

Attempt ALL questions in this section.

Read all questions carefully before attempting.

Use **blue** or **black** ink. Do NOT use gel pens.

Write your answers in the spaces provided. Additional space for answers and rough work is provided at the end of this booklet. If you use this space, write clearly the number of the question you are attempting. Any rough work must be written in this booklet. You should score through your rough work when you have written your fair copy.

SECTION 1— 20 marks

The questions for Section 1 are contained in the booklet Chemistry Section 1—Questions. Read these and record your answers on the grid on Page three opposite.

1.　The answer to each question is **either** A, B, C or D.　Decide what your answer is, then fill in the appropriate bubble (see sample question below).

2.　There is **only one correct** answer to each question.

3.　Any rough working should be done on the additional space for rough working and answers sheet.

Sample Question

To show that the ink in a ball-pen consists of a mixture of dyes, the method of separation would be:

　　A　fractional distillation

　　B　chromatography

　　C　fractional crystallisation

　　D　filtration.

The correct answer is **B**—chromatography.　The answer **B** bubble has been clearly filled in (see below).

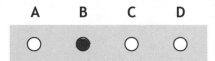

Changing an answer

If you decide to change your answer, cancel your first answer by putting a cross through it (see below) and fill in the answer you want. The answer below has been changed to **D**.

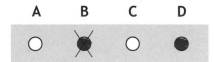

If you then decide to change back to an answer you have already scored out, put a tick (✓) to the **right** of the answer you want, as shown below:

SECTION 1—Answer Grid

	A	B	C	D
1	○	○	○	○
2	○	○	○	○
3	○	○	○	○
4	○	○	○	○
5	○	○	○	○
6	○	○	○	○
7	○	○	○	○
8	○	○	○	○
9	○	○	○	○
10	○	○	○	○
11	○	○	○	○
12	○	○	○	○
13	○	○	○	○
14	○	○	○	○
15	○	○	○	○
16	○	○	○	○
17	○	○	○	○
18	○	○	○	○
19	○	○	○	○
20	○	○	○	○

[BLANK PAGE]

MARKS | DO NOT WRITE IN THIS MARGIN

SECTION 2— 60 marks

Attempt ALL questions.

1. The Eurofighter "Typhoon" is made from many newly developed materials including titanium alloys.

(a) The first step in extracting titanium from its ore is to convert it into titanium(IV) chloride.

Titanium(IV) chloride is a liquid at room temperature and does not conduct electricity.

What type of bonding, does this suggest, is present in titanium(IV) chloride? **1**

(b) Titanium(IV) chloride is then reduced to titanium metal.

The equation for the reaction taking place is:

$$TiCl_4 \ + \ Na \rightarrow \ Ti + \ NaCl$$

(i) Balance the equation. **1**

(ii) What does this reaction suggest about the reactivity of titanium compared to that of sodium? **1**

Total marks 3

MARKS | DO NOT WRITE IN THIS MARGIN

2. (a) Galena is an ore containing lead sulphide, PbS.

   (i) What is the charge on the lead ion in this compound? **1**

   (ii) Calculate the percentage by mass of lead in galena, PbS. **3**

(b) Most metals have to be extracted from their ores.

Place the following metals in the correct space in the table.

copper   mercury   aluminium **1**

You may wish to use the data booklet to help you.

Metal	Method of extraction
	using heat alone
	electrolysis of molten ore
	heating with carbon

Total marks **5**

MARKS | DO NOT WRITE IN THIS MARGIN

3. Cool packs can be used to treat some sports injuries.

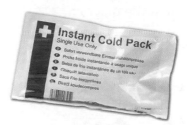

The pack contains solid ammonium nitrate and water in two separate compartments. When the pack is squeezed the ammonium nitrate dissolves in the water forming a solution. This results in a drop in temperature.

(a) The change in temperature in the cool pack can be calculated using the equation below.

$$\text{Temperature change} = \frac{\text{energy change (kJ)}}{\text{mass of water (kg)} \times 4 \cdot 2}$$

Calculate the temperature change using the following information.   2

Energy change (kJ)	6·72
Mass of water (kg)	0·2

(b) Write the ionic formula of ammonium nitrate.   1

Total marks   3

MARKS | DO NOT WRITE IN THIS MARGIN

4. The element carbon can exist in the form of diamond.

    The structure of diamond is shown in the diagram.

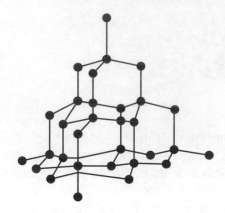

    (a) Name the type of **bonding** and **structure** present in diamond.　　2

    (b) Carbon forms many compounds with other elements such as hydrogen.

         Draw a diagram to show how the outer electrons are arranged in a molecule of methane, CH_4.　　1

Total marks　3

MARKS

5. Volatile organic compounds, VOCs, are organic compounds that can cause damage to the Earth's atmosphere. They may also be harmful or toxic. They are used in paints as solvents and the VOC content is displayed on most paint cans.

LOW
VOC CONTENT 0.30% to 7.99%
VOCs. (Volatile Organic Compounds) contribute to atmospheric pollution.

(a) An example of a VOC compound used in paints is methanal which is the first member of the aldehydes homologous series. Methanal has the structural formula

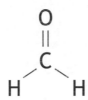

What is meant by the term homologous series? 　1

(b) Methanal is very flammable. A 2 g sample was burned and the heat produced raised the temperature of 200 cm^3 of water from 20.0 °C to 64.7 °C.

Calculate the energy released, in kJ. 　3

You may wish to use the data booklet to help you.

Show your working clearly.

Total marks　4

MARKS | DO NOT WRITE IN THIS MARGIN

6. A student is given the task of identifying the type of bonding and the elements present in an unknown compound.

Using your knowledge of chemistry, describe tests that the student could perform to identify both the bonding and elements present in the unknown compound. The test descriptions should also include examples of possible results and what the results would indicate. 3

Total marks 3

MARKS | DO NOT WRITE IN THIS MARGIN

7. A student was asked to investigate if the type of electrolyte used affects the voltage produced in a cell.

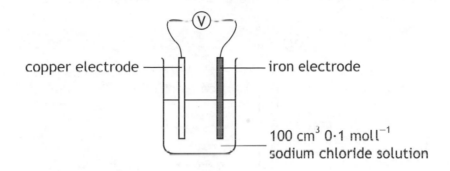

copper electrode ———   ——— iron electrode

100 cm^3 0·1 mol l^{-1}
sodium chloride solution

(a) (i) Complete the labeling of a second cell which could be used to compare the effect of changing the electrolyte from sodium chloride to hydrochloric acid.   **1**

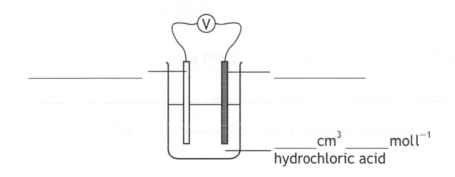

_____ cm^3 _____ mol l^{-1}
hydrochloric acid

(ii) Suggest what should be done to make sure the results are reliable.   **1**

MARKS | DO NOT WRITE IN THIS MARGIN

7. (b) (continued)

(b) Shown is a cell that contains both a metal and a non-metal electrode.

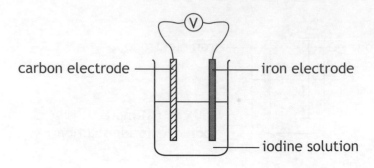

carbon electrode ⎯ ⎯ iron electrode

⎯ iodine solution

(i) The ion-electron equation for the reaction taking place at the carbon electrode is:

$$I_2(aq) \quad + \quad 2e^- \quad \longrightarrow \quad 2I^-(aq)$$

On the diagram, clearly mark the path and direction of electron flow. 1

(ii) What term can be used to describe the reaction taking place at the carbon electrode? 1

Total marks 4

MARKS | DO NOT WRITE IN THIS MARGIN

8. When calcium chloride is dissolved in water, heat is released to the surroundings.

   (a) What term is used to describe chemical reactions which give out heat?   1

   (b) A student investigated how changing the mass of calcium chloride affects the heat released.

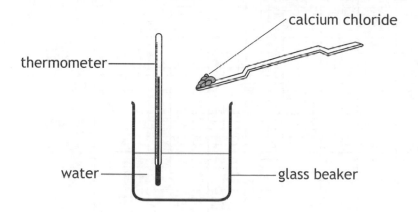

   The results are shown.

Mass of calcium chloride used (g)	Highest temperature reached (°C)
0	20
5	28
10	34
15	41
20	50
25	57

MARKS

8. (b) (continued)

(i) Plot a line graph of these results.   3

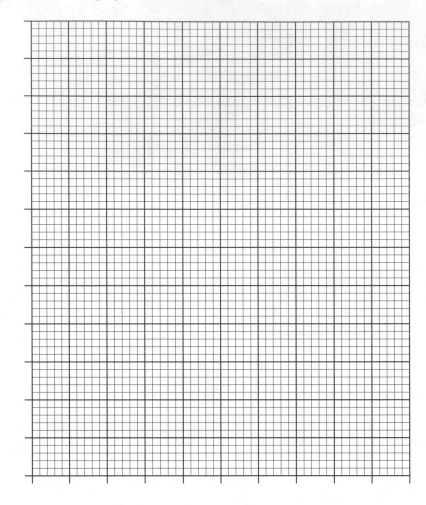

(ii) Using your graph, find the mass of calcium chloride that would give a temperature of 40 °C.   1

(c) Suggest an improvement that could be made to this experiment.   1

Total marks   6

MARKS | DO NOT WRITE IN THIS MARGIN

9. Tritium, $^3_1$H, is an isotope of hydrogen. It is formed in the upper atmosphere when hydrogen atoms capture neutrons from cosmic rays. The tritium atoms then decay by beta-emission.

$$^3_1H \quad \rightarrow \qquad\qquad + \quad ^{\,0}_{-1}e^-$$

(a) Complete the nuclear equation above for the beta-decay of tritium atoms.

1

(b) The concentration of tritium atoms in fallen rainwater is found to decrease over time. The age of any product made with water can be estimated by measuring the concentration of tritium atoms.

In a bottle of wine, the concentration of tritium atoms was found to be 25% of the concentration found in rain.

Given that the half-life of tritium is 12.3 years, how old is the wine?

2

Total marks   3

MARKS

10. Some sterilising pads contain a 65% solution of Isopropyl alcohol in water.

Isopropyl alcohol has the systematic name propan-2-ol.

(a) Draw the full structural formula of isopropyl alcohol. 1

(b) Name an isomer of propan-2-ol. 1

(c) What is the name of the functional group present in all alcohols? 1

(d) Some sterilising pads also contain ethanol. If a typical pad contains 0·46 g of ethanol, how many moles of ethanol does it contain? 2

(e) Alcohols can be used to produce esters. What group of compounds must alcohols react with to produce esters? 1

Total marks 6

MARKS | DO NOT WRITE IN THIS MARGIN

11. Read the passage below and answer the questions that follow.

Scientists Investigate Release of Bromine in Polar Regions

Ozone plays a key role not only in the atmosphere, but also on the ground. While at ground level it is not particularly relevant for the protection from UV radiation, it is for the self-cleaning of the atmosphere and removal of contaminants.

In the 1990s researchers discovered that the extensive ozone depletion in the atmosphere close to the ground in the Arctic and Antarctic was due to a reaction of bromine with ozone (O_3), producing bromine oxide (Br_2O) and oxygen. This bromine is released in autocatalytic processes.

During the polar spring, the resulting bromine oxide clouds can spread over several thousand square kilometers. "It is by far the largest release of bromine on our planet," says Prof. Platt of the Institute of Environmental Physics at Heidelberg University. The precise processes involved are quite complex and are still a topic of current research.

The passage on Bromine in Polar Regions was taken from an article published on www.uni-heidelberg.de.

(a) What role does ozone play in our atmosphere?   1

(b) Write the formula equation for the reaction of bromine with ozone. There is no need to balance the equation.   1

(c) CFCs such as dichlorofluoromethane are also broken down by UV radiation to produce very reactive free radicals such as chlorine atoms. These chlorine atoms react with the ozone as shown in the equation.

$$Cl(g) + O_3(g) \rightarrow ClO(g) + O_2(g)$$

What mass of ozone would react with 71g of chlorine free radicals?   3

Total marks   5

MARKS

12. The concentration of ethanoic acid in vinegar can be calculated by neutralising a sample with 0.5 mol l^{-1} sodium hydroxide solution.

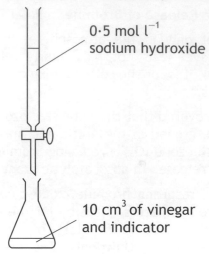

0·5 mol l^{-1}
sodium hydroxide

10 cm^3 of vinegar
and indicator

(a) Draw the full structural formula of ethanoic acid. 1

(b) An average of 20 cm^3 of sodium hydroxide was required to neutralise the vinegar.

The equation for the reaction is

$$CH_3COOH + NaOH \rightarrow CH_3COONa + H_2O$$

Calculate the concentration, in mol l^{-1}, of the ethanoic acid in the vinegar. 3

(c) Ethanoic acid is classed as a weak acid, but an acid such as hydrochloric acid is classed as a strong acid. This means that a 1·0 mol l^{-1} solution of hydrochloric acid has a pH of 0 but a 1·0 mol l^{-1} solution of ethanoic acid has a pH of 2.

Using your knowledge of chemistry, describe several ways in which the strength of these acids could be compared experimentally. 3

Total marks   7

13. Kevlar is a synthetic fibre that can be used to reinforce the walls of tyres.

(a) Kevlar is made from the following monomers.

(i) When these two monomers combine, hydrogen chloride (HCl) is also produced. Draw the structure of the repeating unit formed from these two monomers.

1

(ii) Name the type of polymerisation that takes place to form Kevlar.

1

(b) The polymer shown can be used to produce belts for car engines but increasingly car manufacturers are using Kevlar as a replacement.

The polymer shown is an example of a polyester. Draw the structural formula of the two monomers that were used to produce this polymer.

2

Total marks   4

MARKS DO NOT WRITE IN THIS MARGIN

14. The graph shows how the solubility of potassium chloride changes with temperature.

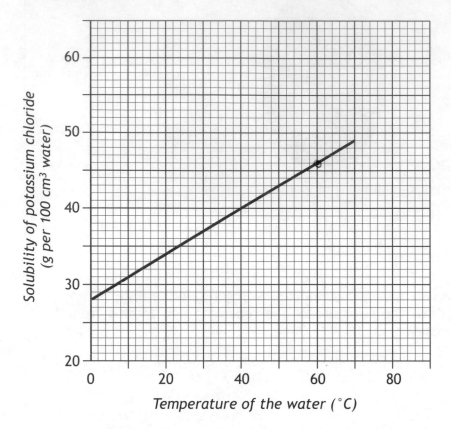

(a) From the graph, what is the maximum mass of potassium chloride that will dissolve at 60 °C?

1

(b) The potassium chloride solution is cooled from 60°C to 30°C. A solid forms at the bottom of the beaker.

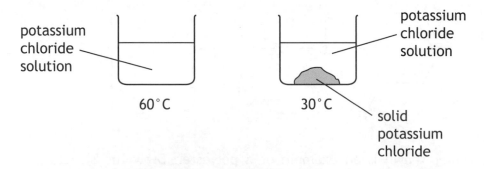

(i) Using the graph, calculate the mass of solid potassium chloride formed at the bottom of the beaker at 30 °C.

2

MARKS

DO NOT WRITE IN THIS MARGIN

14. (b) (continued)

    (ii) What method could be used to separate the solid that forms from the potassium chloride solution?   **1**

Total Marks   4

[END OF MODEL PAPER]

MARKS

ADDITIONAL SPACE FOR ROUGH WORKING AND ANSWERS

ADDITIONAL SPACE FOR ROUGH WORKING AND ANSWERS

ADDITIONAL SPACE FOR ANSWERS

Additional graph paper for Question 8 (b) (i)

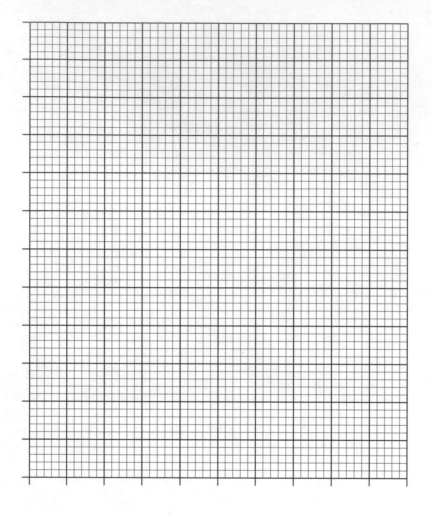

Model Paper 3

Whilst this Model Paper has been specially commissioned by Hodder Gibson for use as practice for the National 5 exams, the key reference documents remain the SQA Specimen Paper 2013 and the SQA Past Papers 2014 and 2015.

HODDER
GIBSON
LEARN MORE

National Qualifications MODEL PAPER 3

Chemistry
Section 1—Questions

Duration — 2 hours

Instructions for completion of Section 1 are given on Page two of the question paper.

Record your answers on the grid on Page three of your answer booklet.

Do not write in this booklet.

Before leaving the examination room you must give your answer booklet to the Invigilator. If you do not, you may lose all the marks for this paper.

HODDER GIBSON
LEARN MORE

SECTION 1

1. Which of the following has metallic bonding?

    A   Calcium

    B   Carbon

    C   Oxygen

    D   Fluorine

2. The table shows information about an ion.

Particle	Number
electron	19
neutron	20
proton	18

    The charge on the ion is

    A   1+

    B   1-

    C   2+

    D   2-.

3. Isotopes of the same element have identical

    A   nuclei

    B   mass number

    C   number of neutrons

    D   number of protons.

4. Which of the following compounds could be used to represent the structure of an ionic compound?

A

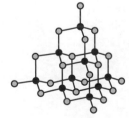

C

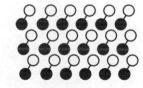

B

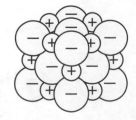

D

5. The course of a reaction was followed by measuring the volume of gas produced over time. 40 cm^3 of gas was collected after 100 s.

    What was the average rate of reaction, in cm^3s^{-1}, over 100 seconds?

    A   0.04

    B   0.4

    C   2.5

    D   4

6. The correct formula of calcium sulfite is

    A CaS

    B $CaSO_3$

    C $CaSO_4$

    D CaS_2O_3.

7. Which line in the table correctly shows the properties of an ionic compound?

	Melting Point (°C)	Conducts Electricity?	
		Solid	Liquid
A	181	Yes	Yes
B	-95	No	No
C	686	No	Yes
D	1700	No	No

8. Compared to pure water, an acidic solution contains

    A   only hydrogen ions

    B   more hydrogen ions than hydroxide ions

    C   more hydroxide ions than hydrogen ions

    D   equal numbers of hydrogen ions and hydroxide ions.

9. What name is given to the reaction shown by the following equation?

    $$C_2H_4 + H_2 \rightarrow C_2H_6$$

    A   Combustion

    B   Neutralisation

    C   Polymerisation

    D   Addition

10. The balanced equation for the complete combustion of a hydrocarbon X is shown below.

$$X(g) + 2O_2(g) \rightarrow CO_2(g) + 2H_2O(l)$$

Which of the following is the correct formula of hydrocarbon X?

A CH_4

B C_2H_6

C C_3H_8

D C_4H_{10}

11. The table shows the result of heating two compounds with acidified potassium dichromate solution.

Compound	Acidified potassium dichromate solution
H H O H │ │ ‖ │ H — C — C — C — C — H │ │ │ H H H	stays orange
H H H O │ │ │ ‖ H — C — C — C — C — H │ │ │ H H H	turns green

Which of the following compounds will **not** turn acidified potassium dichromate solution green?

A
```
    H   O   H
    │   ‖   │
H — C — C — C — H
    │       │
    H       H
```

B
```
    H   H   O
    │   │   ‖
H — C — C — C — H
    │   │
    H   H
```

C
```
    H   O
    │   ‖
H — C — C — H
    │
    H
```

D
```
    O
    ‖
H — C — H
```

12. Which of the following is **not** the first member of a homologous series?

A

$$H - \underset{\underset{H}{|}}{\overset{\overset{H}{|}}{C}} - H$$

B

C

$$\underset{\underset{H}{|}}{\overset{\overset{H}{|}}{C}} = \underset{\underset{H}{|}}{\overset{\overset{H}{|}}{C}}$$

D

$$\underset{\underset{H}{|}}{\overset{\overset{CH_3}{|}}{C}} = \underset{\underset{H}{|}}{\overset{\overset{H}{|}}{C}}$$

13. Which of the following would quickly decolourise bromine solution?

   A C_2H_4

   B C_3H_8

   C C_4H_{10}

   D C_5H_{12}

14. $H^+(aq) + NO_3^-(aq) + K^+(aq) + OH^-(aq) \longrightarrow K^+(aq) + NO_3^-(aq) + H_2O(l)$

The spectator ions in the reaction are

   A $H^+(aq)$ and $K^+(aq)$

   B $NO_3^-(aq)$ and $OH^-(aq)$

   C $H^+(aq)$ and $OH^-(aq)$

   D $K^+(aq)$ and $NO_3^-(aq)$.

15. The following statements relate to four different metals, **P**, **Q**, **R** and **S**.

Metal **P** displaces metal **Q** from a solution containing ions of **Q**.

In a cell, electrons flow from metal **S** to metal **P**.

Metal **R** is the only metal which can be obtained from its ore by heat alone.

The order of reactivity of the metals, starting with the **most** reactive is

A   S, P, Q, R

B   R, Q, P, S

C   R, S, Q, P

D   S, Q, P, R.

16. Some metals can be obtained from their metal oxides by heat alone.

Which of the following oxides would produce a metal when heated?

A   Calcium oxide

B   Copper oxide

C   Zinc oxide

D   Silver oxide

17. Polythene terephthalate (PET) is used to make plastic bottles, which can easily be recycled by heating and reshaping.

A section of the PET structure is shown.

Which line in the table best describes PET?

	Type of polymer	Natural/Synthetic
A	addition	synthetic
B	condensation	natural
C	addition	natural
D	condensation	synthetic

18. Four cells were made by joining copper, iron, magnesium and zinc to silver.

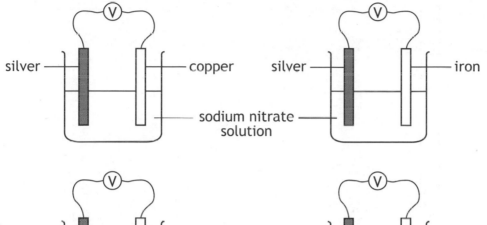

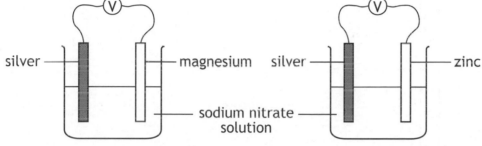

Which of the following will be the voltage of the cell containing silver joined to copper?

You may wish to use the data booklet to help you.

A   0.5 V

B   0.9 V

C   1.1 V

D   2.7 V

19. Which particle will be formed when an atom of $^{212}_{83}Bi$ emits an alpha particle?

A $^{207}_{82}Pb$

B $^{208}_{81}Tl$

C $^{209}_{80}Hg$

D $^{210}_{79}Au$

20. In which of the following test tubes will a reaction occur?

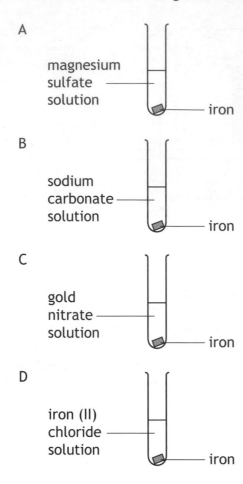

A

magnesium
sulfate
solution — iron

B

sodium
carbonate — iron
solution

C

gold
nitrate — iron
solution

D

iron (II)
chloride — iron
solution

**[END OF SECTION 1. NOW ATTEMPT THE QUESTIONS IN SECTION 2
OF YOUR QUESTION AND ANSWER BOOKLET.]**

National Qualifications MODEL PAPER 3

Chemistry
Section 1—Answer Grid and Section 2

Duration — 2 hours

Total marks — 80

SECTION 1 — 20 marks

Attempt ALL questions in this section.

Instructions for completion of Section 1 are given on Page two.

SECTION 2 — 60 marks

Attempt ALL questions in this section.

Read all questions carefully before attempting.

Use **blue** or **black** ink. Do NOT use gel pens.

Write your answers in the spaces provided. Additional space for answers and rough work is provided at the end of this booklet. If you use this space, write clearly the number of the question you are attempting. Any rough work must be written in this booklet. You should score through your rough work when you have written your fair copy.

SECTION 1— 20 marks

The questions for Section 1 are contained in the booklet Chemistry Section 1—Questions.
Read these and record your answers on the grid on Page three opposite.

1. The answer to each question is **either** A, B, C or D. Decide what your answer is, then fill in the appropriate bubble (see sample question below).

2. There is **only one correct** answer to each question.

3. Any rough working should be done on the additional space for rough working and answers sheet.

Sample Question

To show that the ink in a ball-pen consists of a mixture of dyes, the method of separation would be:

     A   fractional distillation

     B   chromatography

     C   fractional crystallisation

     D   filtration.

The correct answer is **B**—chromatography. The answer **B** bubble has been clearly filled in (see below).

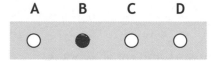

Changing an answer

If you decide to change your answer, cancel your first answer by putting a cross through it (see below) and fill in the answer you want. The answer below has been changed to **D**.

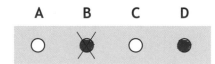

If you then decide to change back to an answer you have already scored out, put a tick (✓) to the **right** of the answer you want, as shown below:

SECTION 1—Answer Grid

	A	B	C	D
1	○	○	○	○
2	○	○	○	○
3	○	○	○	○
4	○	○	○	○
5	○	○	○	○
6	○	○	○	○
7	○	○	○	○
8	○	○	○	○
9	○	○	○	○
10	○	○	○	○
11	○	○	○	○
12	○	○	○	○
13	○	○	○	○
14	○	○	○	○
15	○	○	○	○
16	○	○	○	○
17	○	○	○	○
18	○	○	○	○
19	○	○	○	○
20	○	○	○	○

[BLANK PAGE]

MARKS | DO NOT WRITE IN THIS MARGIN

SECTION 2— 60 marks

Attempt ALL questions.

1. Iron displaces silver from silver(I) nitrate solution.

   The equation for the reaction is:

   $$Fe(s) + 2Ag^+(aq) + 2NO_3^-(aq) \rightarrow Fe^{2+}(aq) + 2Ag(s) + 2NO_3^-(aq)$$

   (a) Write the ion-electron equation for the reduction step in the reaction.   **1**

   You may wish to use the data booklet to help you.

   (b) This reaction can also be carried out in a cell.

   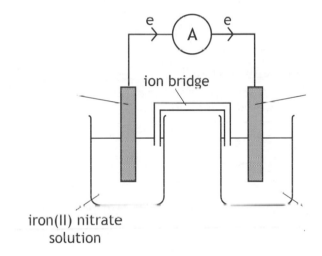

   ion bridge

   iron(II) nitrate
   solution

   Complete the three labels on the diagram.   **1**

   (c) What is the purpose of the ion bridge?   **1**

   Total marks   **3**

MARKS | DO NOT WRITE IN THIS MARGIN

2. In 2011 33% of Scotland's electricity was produced from nuclear power stations. Coal accounted for 21% with renewable and other sources accounting for the rest.

Green salt (uranium tetrafluoride) can be used to produce fuel for nuclear power stations. It is produced from uranium ore.

(a) Uranium can be extracted from green salt in a redox reaction with magnesium metal.

$$Mg + UF_4 \rightarrow MgF_2 + U$$

Balance the equation.   1

(b) This reaction is carried out at temperatures of over 1100°C in an argon atmosphere.

Suggest a reason why the reaction is not carried out in air.

(c)   **Properties of uranium hexafluoride (UF$_6$)**

Appearance	Colourless solid
Density	5.09 g/cm^3
Melting point	64.8°C

Use this information to suggest the type of bonding present in uranium hexafluoride (UF$_6$).   1

(d) When an electric current is passed through water, hydrogen and oxygen are produced. The hydrogen can then be used as a fuel for fuel cells.

Using your knowledge of chemistry, give arguments for and against the suggestion that hydrogen is a pollution free fuel.   3

Total marks   6

MARKS | DO NOT WRITE IN THIS MARGIN

3. Copper(II) sulfate crystals can be prepared by reacting copper carbonate with dilute sulfuric acid. Water is also produced in the reaction.

(a) Using symbols and formulae, write the chemical equation for this reaction.

There is no need to balance the equation.

1

(b) The four steps involved in the preparation of copper(II) sulfate are shown below:

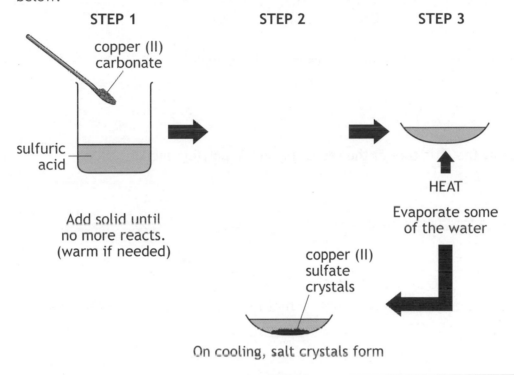

STEP 1

copper (II) carbonate

sulfuric acid

Add solid until no more reacts. (warm if needed)

STEP 2

STEP 3

HEAT

Evaporate some of the water

copper (II) sulfate crystals

On cooling, salt crystals form

STEP 4

Draw the labeled diagram for the 2nd step.

1

(c) In the 'reaction', suggest how you could tell that the reaction is complete.

1

(d) Why must the copper carbonate be added in excess to the sulfuric acid?

1

Total marks   **4**

MARKS | DO NOT WRITE IN THIS MARGIN

4.   Poly(ethenol) is one of the substances used to cover dishwasher tablets.

A section of the poly(ethenol) polymer is shown.

$$-CH_2-CH-CH_2-CH-CH_2-CH-$$
$$\quad\quad\ |\quad\quad\quad\ |\quad\quad\quad\ |$$
$$\quad\quad\ OH\quad\quad\ OH\quad\quad\ OH$$

(a) Name the functional group present in this polymer.   1

(b) Draw the structure of the repeating unit in poly(ethenol).   1

(c) Name the type of polymerisation that takes place to form poly(ethenol).   1

Total marks   3

MARKS | DO NOT WRITE IN THIS MARGIN

5. The diagram shows the apparatus used to prepare chlorine gas.

Concentrated hydrochloric acid is reacted with potassium permanganate.

The gas produced is bubbled through water to remove any unreacted hydrochloric acid and is then dried by bubbling through concentrated sulfuric acid.

(a) Complete the diagram for the preparation of chlorine gas by adding the labels for concentrated sulfuric acid, potassium permanganate and water. **1**

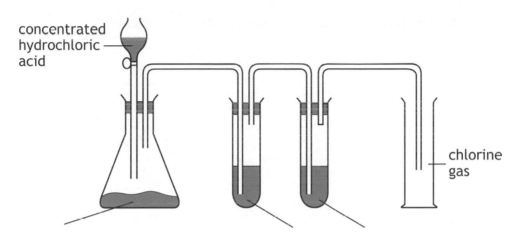

(b) Chlorine is a member of the Group 7 elements.

The graph shows the melting points of these elements.

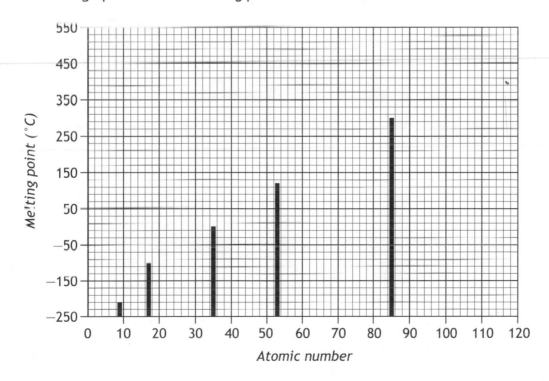

MARKS DO NOT WRITE IN THIS MARGIN

5. (b) (continued)

(i) State the relationship between the atomic number and the melting point of the Group 7 elements. **1**

(ii) The next member of this group would have an atomic number of 117.

Using the graph, predict the melting point of this element. **1**

Total marks 3

MARKS | DO NOT WRITE IN THIS MARGIN

6. Alkynes are a homologous series of hydrocarbons, which contain carbon to carbon triple bonds. Two members of this series are shown.

$$H - C \equiv C - \overset{\displaystyle \overset{H}{|}}{\underset{\displaystyle \underset{H}{|}}{C}} - \overset{\displaystyle \overset{H}{|}}{\underset{\displaystyle \underset{H}{|}}{C}} - H \qquad H - C \equiv C - \overset{\displaystyle \overset{H}{|}}{\underset{\displaystyle \underset{H}{|}}{C}} - \overset{\displaystyle \overset{H}{|}}{\underset{\displaystyle \underset{H}{|}}{C}} - \overset{\displaystyle \overset{H}{|}}{\underset{\displaystyle \underset{H}{|}}{C}} - H$$

butyne   pentyne

(a) Name the first member of this series.   1

(b) Alkynes can be prepared by reacting a dibromoalkane with potassium hydroxide solution.

$$H - \overset{\displaystyle \overset{H}{|}}{\underset{\displaystyle \underset{Br}{|}}{C}} - \overset{\displaystyle \overset{H}{|}}{\underset{\displaystyle \underset{Br}{|}}{C}} - \overset{\displaystyle \overset{H}{|}}{\underset{\displaystyle \underset{H}{|}}{C}} - H + 2KOH \rightarrow H - C \equiv C - \overset{\displaystyle \overset{H}{|}}{\underset{\displaystyle \underset{H}{|}}{C}} - H + 2KBr + H_2O$$

dibromoalkane   propyne

(i) Draw a structural formula for the alkyne formed when the dibromoalkane shown below reacts with potassium hydroxide solution.   1

$$H - \overset{\displaystyle \overset{H}{|}}{\underset{\displaystyle \underset{H}{|}}{C}} - \overset{\displaystyle \overset{H}{|}}{\underset{\displaystyle \underset{Br}{|}}{C}} - \overset{\displaystyle \overset{H}{|}}{\underset{\displaystyle \underset{Br}{|}}{C}} - \overset{\displaystyle \overset{H}{|}}{\underset{\displaystyle \underset{H}{|}}{C}} - H + 2KOH \rightarrow$$

(ii) Suggest a reason why the dibromoalkane shown below does not form an alkyne when it is added to potassium hydroxide solution.   1

$$H - \overset{\displaystyle \overset{H}{|}}{\underset{\displaystyle \underset{H}{|}}{C}} - \overset{\displaystyle \overset{H}{|}}{\underset{\displaystyle \underset{Br}{|}}{C}} - \overset{\displaystyle \overset{H}{|}}{\underset{\displaystyle \underset{H}{|}}{C}} - \overset{\displaystyle \overset{H}{|}}{\underset{\displaystyle \underset{Br}{|}}{C}} - \overset{\displaystyle \overset{H}{|}}{\underset{\displaystyle \underset{H}{|}}{C}} - H$$

Total marks   3

MARKS | DO NOT WRITE IN THIS MARGIN

7. Gold is a very soft metal. In order to make it harder, goldsmiths mix it with silver. The quality of the gold is indicated in carats.

(a) The graph shows information about the quality of gold.

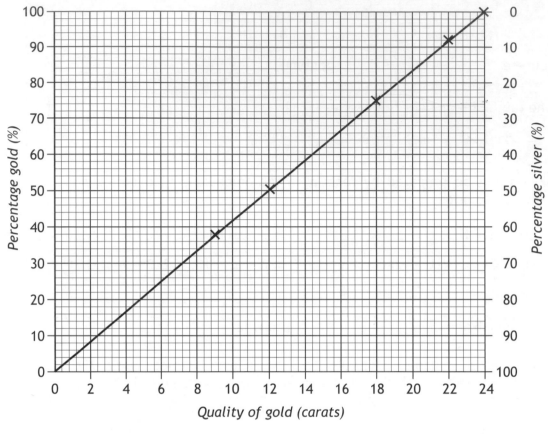

(i) What is the percentage of silver in an 18-carat gold ring?   **1**

(ii) Calculate the mass of silver in an 18-carat gold ring weighing 6 g.   **1**

(b) Silver tarnishes in the presence of hydrogen sulfide forming black silver sulfide, Ag_2S.

The equation for the reaction is:

$$4Ag + 2H_2S + O_2 \longrightarrow 2Ag_2S + 2H_2O$$

What mass of silver sulfide would be formed from 1·08 g of silver?   **3**

Total marks   5

MARKS | DO NOT WRITE IN THIS MARGIN

8. Chemicals in food provide flavour and smell. Ketones are responsible for the flavour in blue cheese.

Two examples of ketones are shown below.

```
    H   O   H   H   H                           H   H   O   H   H
    |   ||  |   |   |                           |   |   ||  |   |
H — C — C — C — C — C — H              H — C — C — C — C — C — H
    |       |   |   |                           |   |       |   |
    H       H   H   H                           H   H       H   H
```

pentan-2-one   pentan-3-one

(a) Draw a structural formula for hexan-3-one.   1

(b) Suggest a name for the ketone shown.   1

```
    H   H   H   O   H   H   H
    |   |   |   ||  |   |   |
H — C — C — C — C — C — C — C — H
    |   |   |       |   |   |
    H   H   H       H   H   H
```

(c) Information about the boiling points of four ketones is shown in the table.

Ketone	Boiling point (°C)
C_3H_6O	56
C_4H_8O	80
$C_5H_{10}O$	102
$C_6H_{12}O$	127

Predict the boiling point of $C_7H_{14}O$.   1

(d) Sweets, such as pineapple cubes, contain the ester methyl butanoate to provide flavour.

  (i) Give another use of esters.   1

  (ii) Methyl butanoate can be broken down to form methanol and butanoic acid.
       Draw the full structural formula of butanoic acid.   1

Total marks   5

9. Alpha, beta and gamma radiation is passed from a source through an electric field. The gamma radiation passes directly through, unaffected by the charged plates.

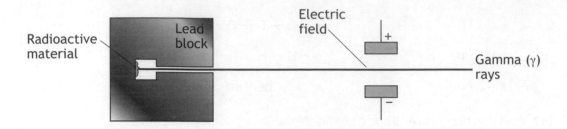

Draw lines on the diagram to show what effect you would expect the charged plates would have on alpha and beta particles. Remember to label each of your lines.   **2**

Total marks   2

MARKS | DO NOT WRITE IN THIS MARGIN

10. Read the following passage carefully and answer the questions that follow.

Self Distilling Vodka

Scientists were investigating the permeability of a material called graphene oxide. This is graphene that has been reacted with a strong oxidising agent, making it more soluble and easier to deal with. They created membranes made up of small pieces of graphene oxide which pile up like bricks to form an interlocked structure, and then tested how gas-proof they were by using the film as a lid for a container full of various gases.

They found that despite being 500 times thinner than a human hair, it completely stopped the gases hydrogen, nitrogen and argon from escaping.

It even stopped helium which, being a tiny single atom will escape from party balloons very quickly, and can even diffuse out through a millimetre of glass. They then tried various liquids, and found similar behaviour for ethanol, hexane, acetone, decane and propanol vapour, but when they tried normal water it behaved as if the membrane wasn't there, escaping at least a hundred thousand times faster than any of the other materials. They think the water is forming a layer one molecule thick between the layers of graphene, blocking the route for everything else, but if it dries out, this gap shrinks and seals up. To make use of this behaviour they put some vodka in the container, and left it for a few days. Normally ethanol evaporates faster than water so vodka gets weaker over time, but with their membrane, which blocked the ethanol, the vodka got stronger and stronger.

Taken from the article "Self Distilling Vodka" by Dave Ansell, published on thenakedscientists.com January 2012.

(a) Name one gas prevented from escaping by the graphene oxide?   1

(b) Why is helium found as a single atom?   1

(c) How much faster was water able to escape compared to the other liquids tested?   1

(d) "Propanol vapour was also unable to escape through the graphene oxide."

Give the correct **systematic name** of the two isomers of propanol that may have been used.   2

Total marks   5

MARKS

DO NOT WRITE IN THIS MARGIN

11. Glass is made from the chemical silica, SiO_2, which is covalently bonded and has a melting point of 1700 °C.

(a) What does the melting point of silica suggest about its structure?   1

(b) Antimony(III) oxide is added to reduce any bubbles that may appear during the manufacturing process.

Write the chemical formula for antimony(III) oxide.   1

(c) In the manufacture of glass, other chemicals can be added to alter the properties of the glass. The element boron can be added to glass to make ovenproof dishes.

(i) Information about an atom of boron is given in the table below.

Particle	Number
proton	5
neutron	6

Use this information to complete the nuclide notation for this atom of boron.

B   1

(ii) Atoms of boron exist which have the same number of protons but a different number of neutrons.

What name can be used to describe these different types of boron atoms?   1

Total marks   4

MARKS | DO NOT WRITE IN THIS MARGIN

12. The flow diagram shows how ammonia is converted to nitric acid.

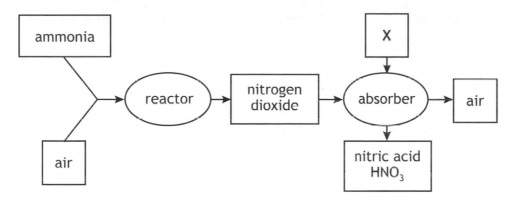

(a) Name the industrial process used to manufacture ammonia.   1

(b) Name substance **X**.   1

(c) Ammonia and nitric acid react together to form ammonium nitrate, NH_4NO_3.
     Calculate the percentage by mass of nitrogen in ammonium nitrate.   3
     Show your working clearly.

Total marks   5

MARKS

13. 'Mag Wheels' were a popular type of alloy wheel fitted to sports cars in the 1950s and 1960s. The wheels were produced from an alloy of magnesium and were favoured over wheels made of steel, which is an alloy of iron by manufacturers of sport cars. Their popularity faded and most alloys wheels fitted to sport cars are now made from alloys of aluminium.

Using your knowledge of chemistry, give reasons for and against the use of alloys of iron, magnesium and aluminium in the production of alloy wheels.

3

Total marks 3

MARKS

<small>DO NOT WRITE IN THIS MARGIN</small>

14. Research is being carried out into making chemicals that can be used to help relieve the side effects of chemotherapy.

One of the reactions in this process is shown

Chemical A + hydrogen   →   Chemical B

(a) As this reaction proceeds, the hydrogen is used up which results in a decrease in pressure.

Time (min)	0	5	10	15	20	30	35	45
Decrease in pressure (bar)	0	0·6	1·2	1·7	2·2	2·9	3·1	3·1

Draw a line graph showing the decrease in pressure as time proceeds.   3

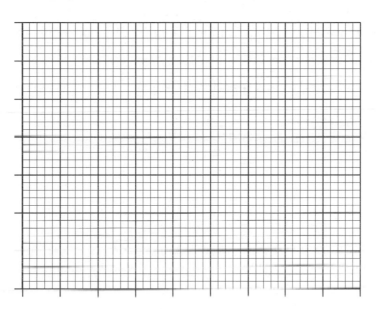

(b) (i) What time did the reaction finish?   1

(ii) Calculate the average rate of the reaction, in bar min⁻¹, between 10 and 20 minutes.   2

Total marks   6

15. The experiment shown can be carried out to establish how much energy is released when ethanol burns.

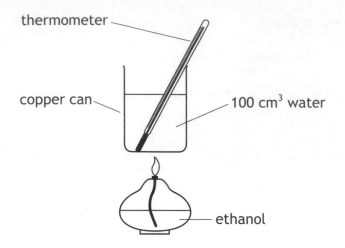

thermometer

copper can

$100 \ cm^3$ water

ethanol

In the experiement it was found that burning 0.1g of ethanol increased the temperature of the water by 7°C.

Calculate the energy released in this reaction, in kJ.   3

Show your working clearly.

Total marks   3

[END OF MODEL PAPER]

ADDITIONAL SPACE FOR ROUGH WORKING AND ANSWERS

ADDITIONAL SPACE FOR ROUGH WORKING AND ANSWERS

MARKS | DO NOT WRITE IN THIS MARGIN

ADDITIONAL SPACE FOR ANSWERS

Additional graph paper for Question 14 (a)

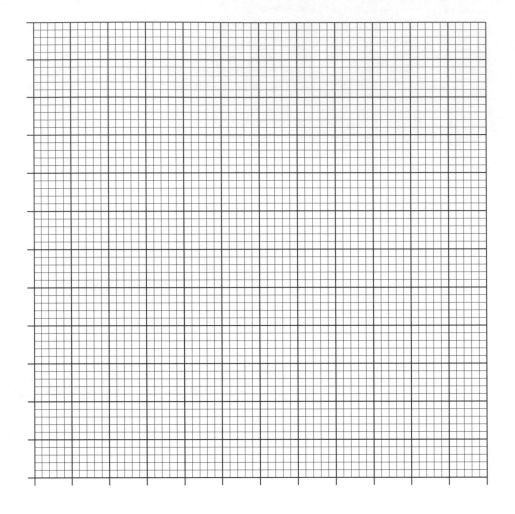

NATIONAL 5

2014

National Qualifications 2014

X713/75/02

Chemistry
Section 1—Questions

MONDAY, 12 MAY

9:00 AM–11:00 AM

Necessary data will be found in the Chemistry Data Booklet for National 5.

Instructions for the completion of Section 1 are given on Page two of your question and answer booklet X713/75/01.

Record your answers on the answer grid on Page three of your question and answer booklet

Before leaving the examination room you must give your question and answer booklet to the Invigilator; if you do not, you may lose all the marks for this paper.

SECTION 1

1. In a reaction, 60 cm³ of hydrogen gas was collected in 20 s.

   What is the average rate of reaction, in $cm^3 s^{-1}$, over this time?

   A $\dfrac{60}{20}$

   B $\dfrac{20}{\square 0}$

   C $\dfrac{1}{\square\square}$

   D $\dfrac{1}{20}$

2. Molecules in which four different atoms are attached to a carbon atom are said to be chiral.

   Which of the following molecules is chiral?

   A
   ```
          Br
          |
          C
        /  |  \
      H    |    H
          Cl
   ```

   B
   ```
          Br
          |
          C
        /  |  \
      H    |    H
          H
   ```

   C
   ```
          I
          |
          C
        /  |  \
     Cl    |    H
          H
   ```

   D
   ```
          I
          |
          C
        /  |  \
      H    |    Br
          F
   ```

Page two

3. What is the charge on the zinc ion in the compound zinc phosphate $Zn_3(PO_4)_2$?

    A   2+

    B   3+

    C   2−

    D   3−

4. Fe_2O_3 + x CO $\longrightarrow$ y Fe   + $3CO_2$

    This equation will be balanced when

    A $x = 1$ and $y = 2$

    B $x = 2$ and $y = 2$

    C $x = 3$ and $y = 2$

    D $x = 2$ and $y = 3$.

5. An acidic solution contains

    A   only hydrogen ions

    B   only hydroxide ions

    C   more hydrogen ions than hydroxide ions

    D   more hydroxide ions than hydrogen ions.

6. Which of the following oxides, when shaken with water, would give an alkaline solution?

    A   Calcium oxide

    B   Nickel oxide

    C   Nitrogen dioxide

    D   Sulfur dioxide

7. Which of the following compounds is **not** a salt?

    A   Calcium nitrate

    B   Sodium chloride

    C   Potassium sulfate

    D   Magnesium hydroxide

8. $H^+(aq) + NO_3^-(aq) + K^+(aq) + OH^-(aq) \longrightarrow K^+(aq) + NO_3^-(aq) + H_2O(\ell)$

   The spectator ions present in the reaction above are

   A $K^+(aq)$ and $NO_3^-(aq)$

   B $K^+(aq)$ and $H^+(aq)$

   C $OH^-(aq)$ and $NO_3^-(aq)$

   D $H^+(aq)$ and $OH^-(aq)$.

9. The molecular formula for cyclohexane is

   A C_6H_6

   B C_6H_{10}

   C C_6H_{12}

   D C_6H_{14}.

10.

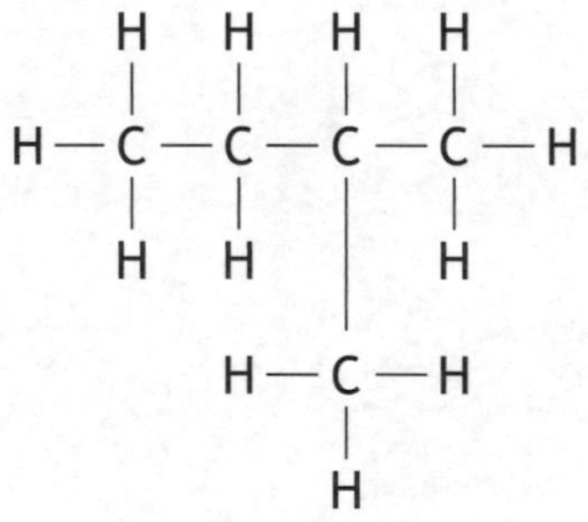

   The systematic name for the structure shown is

   A   1,1-dimethylpropane

   B   2-methylbutane

   C   3-methylbutane

   D   2-methylpentane.

11. Petrol is a mixture of hydrocarbons.

The tendency of a hydrocarbon to ignite spontaneously is measured by its octane number.

	Hydrocarbon	Octane number
1	3-methylpentane	74·5
2	butane	93·6
3	pentane	61·7
4	2-methylpentane	73·4
5	hexane	24·8
6	methylcyclopentane	91·3

A student made the hypothesis that as the chain length of a hydrocarbon increases, the octane number decreases.

Which set of three hydrocarbons should have their octane numbers compared in order to test this hypothesis?

A   1, 4, 6

B   1, 2, 4

C   2, 3, 5

D   3, 4, 5

12. Propene reacts with hydrogen bromide to form two products.

$$
\begin{array}{c}
\quad\; H \\
\quad\; | \\
H-C-C=C-H \\
\quad\; |\quad |\quad | \\
\quad\; H\quad H\quad H
\end{array}
\xrightarrow{\ \text{HBr}\ }
\begin{array}{c}
H\quad Br\quad H \\
|\quad\; |\quad\; | \\
H-C-C-C-H \\
|\quad\; |\quad\; | \\
H\quad H\quad H \\[2ex]
H\quad H\quad Br \\
|\quad\; |\quad\; | \\
H-C-C-C-H \\
|\quad\; |\quad\; | \\
H\quad H\quad H
\end{array}
$$

Which of the following alkenes does **not** form two products on reaction with hydrogen bromide?

A   But-1-ene

B   But-2-ene

C   Pent-1-ene

D   Pent-2-ene

13. Which of the following alcohols has the highest boiling point?

    You may wish to use your data booklet to help you.

    A   Propan-1-ol

    B   Propan-2-ol

    C   Butan-1-ol

    D   Butan-2-ol

14. A reaction is endothermic if

    A   energy is required to start the reaction

    B   heat is released during the reaction

    C   the temperature drops during the reaction

    D   the temperature rises during the reaction.

15. Which of the following metals will **not** react with a dilute solution of hydrochloric acid?

    A   Copper

    B   Iron

    C   Magnesium

    D   Zinc

16. Which metal can be extracted from its oxide by heat alone?

    A   Tin

    B   Zinc

    C   Lead

    D   Silver

17. The ion-electron equations for the oxidation and reduction steps in the reaction between **sulfite ions** and **iron(III) ions** are given below.

oxidation $H_2O(\ell) + SO_3^{2-}(aq) \longrightarrow SO_4^{2-}(aq) + 2H^+(aq) + 2e^-$

reduction $Fe^{3+}(aq) + e^- \longrightarrow Fe^{2+}(aq)$

The redox equation for the overall reaction is

A $H_2O(\ell) + SO_3^{2-}(aq) + Fe^{3+}(aq) \longrightarrow SO_4^{2-}(aq) + 2H^+(aq) + Fe^{2+}(aq) + e^-$

B $H_2O(\ell) + SO_3^{2-}(aq) + 2Fe^{3+}(aq) \longrightarrow SO_4^{2-}(aq) + 2H^+(aq) + 2Fe^{2+}(aq)$

C $SO_4^{2-}(aq) + 2H^+(aq) + Fe^{2+}(aq) + e^- \longrightarrow H_2O(\ell) + SO_3^{2-}(aq) + Fe^{3+}(aq)$

D $SO_4^{2-}(aq) + 2H^+(aq) + 2Fe^{2+}(aq) \longrightarrow H_2O(\ell) + SO_3^{2-}(aq) + 2Fe^{3+}(aq).$

18. The apparatus below was set up.

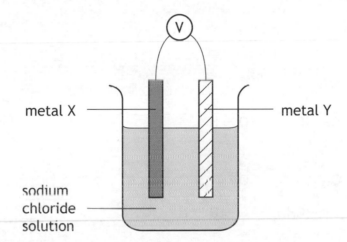

metal X   metal Y

sodium
chloride
solution

Which of the following pairs of metals would give the highest reading on the voltmeter?

	Metal X	Metal Y
A	Iron	Zinc
B	Magnesium	Silver
C	Zinc	Copper
D	Zinc	Silver

[Turn over

19. A section of a condensation polymer is shown below.

$$-\overset{\overset{\displaystyle O}{\|}}{C}-C_6H_4-\overset{\overset{\displaystyle O}{\|}}{C}-O-(CH_2)_2-O-\overset{\overset{\displaystyle O}{\|}}{C}-C_6H_4-\overset{\overset{\displaystyle O}{\|}}{C}-$$

One of the monomers is

$$H-O-\overset{\overset{\displaystyle O}{\|}}{C}-C_6H_4-\overset{\overset{\displaystyle O}{\|}}{C}-O-H$$

The structural formula for the other monomer is

A $\quad H-\overset{\overset{\displaystyle O}{\|}}{C}-O-(CH_2)_2-O-\overset{\overset{\displaystyle O}{\|}}{C}-H$

B $\quad H-O-(CH_2)_2-O-H$

C $\quad H-O-\overset{\overset{\displaystyle O}{\|}}{C}-(CH_2)_2-O-H$

D $\quad H-O-\overset{\overset{\displaystyle O}{\|}}{C}-(CH_2)_2-\overset{\overset{\displaystyle O}{\|}}{C}-O-H$

20. $Ba^{2+}(aq) + 2NO_3^-(aq) + 2Na^+(aq) + SO_4^{2-}(aq) \longrightarrow Ba^{2+}SO_4^{2-}(s) + 2Na^+(aq) + 2NO_3^-(aq)$

The type of reaction represented by the equation above is

A   addition

B   displacement

C   neutralisation

D   precipitation.

**[END OF SECTION 1. NOW ATTEMPT THE QUESTIONS IN SECTION 2
OF YOUR QUESTION AND ANSWER BOOKLET]**

N5

National Qualifications 2014

Mark

X713/75/01

Chemistry
Section 1—Answer Grid
And Section 2

MONDAY, 12 MAY

9:00 AM–11:00 AM

Fill in these boxes and read what is printed below.

Full name of centre

Town

Forename(s)

Surname

Number of seat

Date of birth

Day   Month   Year

D D   M M   Y Y

Scottish candidate number

Necessary Data will be found in the Chemistry Data Booklet for National 5.

Total marks—80

SECTION 1—20 marks

Attempt ALL questions in this section.

Instructions for the completion of Section 1 are given on Page two.

SECTION 2—60 marks

Attempt ALL questions in this section.

Write your answers clearly in the spaces provided in this booklet. Additional space for answers and rough work is provided at the end of this booklet. If you use this space you must clearly identify the question number you are attempting. Any rough work must be written in this booklet. You should score through your rough work when you have written your final copy.

Use **blue** or **black** ink.

Before leaving the examination room you must give this booklet to the Invigilator; if you do not, you may lose all the marks for this paper.

SECTION 1—20 marks

The questions for Section 1 are contained in the question paper X713/75/02.
Read these and record your answers on the answer grid on Page three opposite.
Do NOT use gel pens.

1. The answer to each question is **either** A, B, C or D. Decide what your answer is, then fill in the appropriate bubble (see sample question below).

2. There is **only one correct** answer to each question.

3. Any rough work must be written in the additional space for answers and rough work at the end of this booklet.

Sample Question

To show that the ink in a ball-pen consists of a mixture of dyes, the method of separation would be

   A   fractional distillation

   B   chromatography

   C   fractional crystallisation

   D   filtration.

The correct answer is **B**—chromatography. The answer **B** bubble has been clearly filled in (see below).

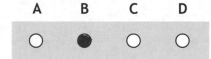

Changing an answer

If you decide to change your answer, cancel your first answer by putting a cross through it (see below) and fill in the answer you want. The answer below has been changed to **D**.

If you then decide to change back to an answer you have already scored out, put a tick (✓) to the **right** of the answer you want, as shown below:

SECTION 1 — Answer Grid

	A	B	C	D
1	○	○	○	○
2	○	○	○	○
3	○	○	○	○
4	○	○	○	○
5	○	○	○	○
6	○	○	○	○
7	○	○	○	○
8	○	○	○	○
9	○	○	○	○
10	○	○	○	○
11	○	○	○	○
12	○	○	○	○
13	○	○	○	○
14	○	○	○	○
15	○	○	○	○
16	○	○	○	○
17	○	○	○	○
18	○	○	○	○
19	○	○	○	○
20	○	○	○	○

[BLANK PAGE]

DO NOT WRITE ON THIS PAGE

[Turn over for Question 1 on *Page six*

DO NOT WRITE ON THIS PAGE

SECTION 2—60 marks
Attempt ALL questions

MARKS

1. In 1911, Ernest Rutherford carried out an experiment to confirm the structure of the atom. In this experiment, he fired positive particles at a very thin layer of gold foil. Most of the particles passed straight through but a small number of the positively charged particles were deflected.

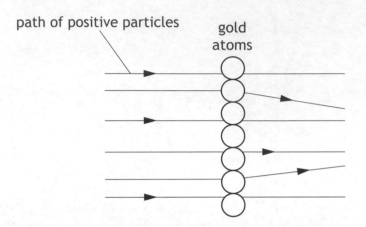

path of positive particles

gold atoms

(a) What caused some of the positive particles to be deflected in this experiment?

1

(b) Gold is the heaviest element to have only one naturally occurring isotope.

The isotope has a mass number of 197.

(i) Complete the table to show the number of each type of particle in this gold atom.

1

You may wish to use the data booklet to help you.

Particle	Number
Proton	
Electron	
Neutron	

(ii) Most elements have more than one isotope.

State what is meant by the term isotope.

1

Total marks   3

2. (a) The properties of a substance depend on its type of bonding and structure. There are four types of bonding and structure.

Discrete covalent molecular	Covalent network	Ionic lattice	Metallic lattice

Complete the table to match up each type of bonding and structure with its properties.

Type of bonding and structure	Properties
	do not conduct electricity and have high melting points
	have high melting points and conduct electricity when liquid but not when solid
	conduct electricity when solid and have a wide range of melting points
	do not conduct electricity and have low melting points

2

(b) Graphene is a substance made of a single layer of carbon atoms.

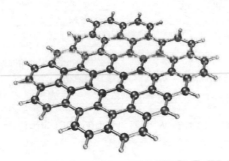

Graphene can conduct electricity.

Suggest what this indicates about some of the electrons in graphene.

1

Total marks   3

MARKS | DO NOT WRITE IN THIS MARGIN

3. Read the passage below and answer the questions that follow.

Potassium – The Super Element

Potassium is an essential element for almost all living things. The human body requires a regular intake of potassium because humans have no mechanism for storing it. Foods rich in potassium include raisins and almonds. Raisins contain 0·86 g of potassium in every 100 g.

Naturally occurring salts of potassium such as saltpetre (potassium nitrate) and potash (potassium carbonate) have been known for centuries. Potassium salts are used as fertilisers.

Potassium was first isolated by Humphry Davy in 1807. Davy observed that when potassium was added to water it formed globules which skimmed about on the surface, burning with a coloured flame and forming an alkaline solution.

(a) State why the human body requires a regular intake of potassium. **1**

(b) Calculate the number of moles of potassium in 100 g of raisins. **2**

Show your working clearly.

(c) State the colour of the flame which would be seen when potassium burns. **1**

You may wish to use the data booklet to help you.

(d) Write the **ionic** formula for saltpetre. **1**

Total marks **5**

MARKS | DO NOT WRITE IN THIS MARGIN

4. Poly(vinylcarbazole) is a plastic which conducts electricity when exposed to light.

The structure of the monomer used to make poly(vinylcarbazole) is

$$
\begin{array}{ccc}
NC_{12}H_8 & & H \\
| & & | \\
C & \!\!=\!\!= & C \\
| & & | \\
H & & H
\end{array}
$$

(a) Draw a section of the polymer showing three monomer units joined together.   **1**

(b) Name the type of polymerisation taking place when these monomers join together.   **1**

Total marks   **2**

[Turn over

MARKS | DO NOT WRITE IN THIS MARGIN

5. Different types of radiation have different penetrating properties.

An investigation was carried out using three radioactive sources.

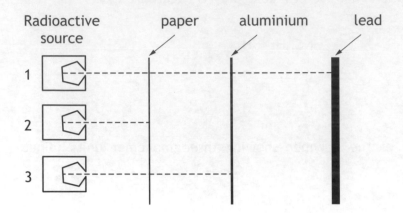

(a) Name the type of radiation emitted by source **2**.   1

(b) The half-life of source **3** is 8 days.

Calculate the fraction of source **3** that would remain after 16 days.   2

Show your working clearly.

(c) Radioisotopes can be made by scientists.

The nuclear equation shows how a radioisotope of element **X** can be made from aluminium.

$$^{27}_{13}\text{Al} \quad + \quad ^{1}_{0}\text{n} \quad \longrightarrow \quad X \quad + \quad ^{4}_{2}\text{He}$$

Name element **X**.   1

Total marks   4

MARKS | DO NOT WRITE IN THIS MARGIN

6. A student reacted acidified potassium permanganate solution with oxalic acid, $C_2H_2O_4$.

$$2MnO_4^-(aq) + 5C_2H_2O_4(aq) + 6H^+(aq) \longrightarrow 2Mn^{2+}(aq) + 10CO_2(g) + 8H_2O(\ell)$$

Using your knowledge of chemistry, comment on how the student could have determined the rate of the reaction.   3

[Turn over

MARKS | DO NOT WRITE IN THIS MARGIN

7. The manufacture of potassium nitrate, for use in fertilisers, can be split into three stages.

(a) (i) In stage **1**, ammonia is produced.

Name the industrial process used to manufacture ammonia.   **1**

(ii) Draw a diagram to show how **all** the outer electrons are arranged in a molecule of ammonia, NH_3.   **1**

(b) In stage **2**, ammonia is converted into nitric acid, HNO_3, as shown in the flow diagram.

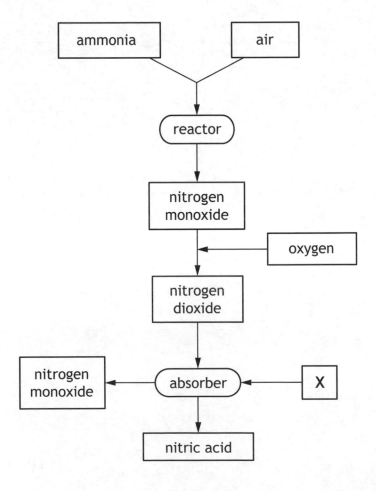

MARKS | DO NOT WRITE IN THIS MARGIN

7. (b) (continued)

(i) Name substance X.   **1**

(ii) **On the flow diagram**, draw an arrow to show how the process can be made more economical.   **1**

(c) In stage **3**, nitric acid is converted to potassium nitrate.

The equation for the reaction taking place is

$$HNO_3(aq) \quad + \quad KOH(aq) \quad \longrightarrow \quad KNO_3(aq) \quad + \quad H_2O(\ell)$$

(i) Name the type of chemical reaction taking place in stage **3**.   **1**

(ii) State how a sample of **solid** potassium nitrate could be obtained from the potassium nitrate solution.   **1**

Total marks   6

[Turn over

MARKS

8. Pheromones are chemicals, produced by living things, that trigger a response in members of the same species.

   When a bee stings an animal the bee also releases a pheromone containing the ester below.

   (a) State another use for esters.   1

   (b) A student made the ester above using ethanoic acid and the following alcohol.

   (i) Name the functional group present in this alcohol.   1

   (ii) Draw a structural formula for an isomer of this alcohol.   1

MARKS | DO NOT WRITE IN THIS MARGIN

8. (b) (continued)

(iii) Ethanoic acid is the second member of a family of compounds which contain the carboxyl functional group.

The full structural formulae for the first three members of this family are shown.

methanoic acid   ethanoic acid   propanoic acid

Suggest a general formula for this family of compounds.   1

(c) The table gives information on some other esters.

Alcohol	Carboxylic acid	Ester
methanol	ethanoic acid	methyl ethanoate
propanol	methanoic acid	propyl methanoate
butanol	ethanoic acid	butyl ethanoate
pentanol	butanoic acid	pentyl butanoate
X	Y	ethyl propanoate

Name X and Y.   2

Total marks   6

MARKS | DO NOT WRITE IN THIS MARGIN

9. Liquefied petroleum gas (LPG), which can be used as a fuel for heating, is a mixture of propane and butane.

(a) Propane and butane are members of the homologous series of alkanes.

Tick (✓) the **two** boxes that correctly describe members of the same homologous series.

1

	Tick (✓)
They have similar chemical properties.	
They have the same molecular formula.	
They have the same general formula.	
They have the same physical properties.	
They have the same formula mass.	

(b) The table gives some information about propane and butane.

Alkane	Boiling Point (°C)
propane	−42
butane	−1

Explain why butane has a higher boiling point than propane.

2

MARKS | DO NOT WRITE IN THIS MARGIN

9.   (continued)

(c)   25 kg of water at 10 °C is heated by burning some LPG.

Calculate the energy, in kJ, required to increase the temperature of the water to 30 °C.   **3**

You may wish to use the data booklet to help you.

Show your working clearly.

(d)   LPG is odourless. In order to detect gas leaks, ethyl mercaptan, C_2H_6S, a smelly gas, is added in small quantities to the LPG mixture.

Suggest one disadvantage of adding sulfur compounds, such as ethyl mercaptan, to fuels such as LPG.   **1**

Total marks   7

MARKS | DO NOT WRITE IN THIS MARGIN

10. The lowest temperature at which a hydrocarbon ignites is called its flash point.

Hydrocarbon	Flash point (°C)
hexane	−23
heptane	−4
octane	13
nonane	31

(a) (i) Using the information in the table, make a general statement linking the flash point to the number of carbon atoms.   **1**

(ii) Predict the flash point, in °C, of decane, $C_{10}H_{22}$.   **1**

MARKS | DO NOT WRITE IN THIS MARGIN

10. (continued)

(b) Nonane burns to produce carbon dioxide and water.

$$C_9H_{20} + 14O_2 \longrightarrow 9CO_2 + 10H_2O$$

Calculate the mass, in grams, of carbon dioxide produced when 32 g of nonane is burned.

3

Show your working clearly.

Total marks   5

[Turn over

11. Chlorine can be produced commercially from concentrated sodium chloride solution in a membrane cell. Only sodium ions can pass through the membrane. These ions move in the direction shown in the diagram.

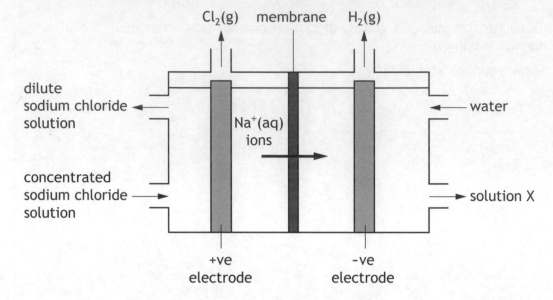

$Cl_2(g)$ membrane $H_2(g)$

dilute sodium chloride solution

water

$Na^+(aq)$ ions

concentrated sodium chloride solution

solution X

+ve electrode

-ve electrode

(a) Write the ion-electron equation for the change taking place at the positive electrode. **1**

You may wish to use the data booklet to help you.

(b) (i) Name solution **X**. **1**

    (ii) The hydrogen gas produced, at the negative electrode, can be used as a fuel.

        Suggest an advantage of using hydrogen as a fuel. **1**

MARKS | DO NOT WRITE IN THIS MARGIN

11. (continued)

(c) The chlorine gas produced can be used to make phosgene, $COCl_2$. Phosgene is used in the manufacture of drugs and plastics.

Draw a possible structure for phosgene.   1

Total marks   4

[Turn over

MARKS

12. Ores are naturally occurring compounds from which metals can be extracted.

(a) When a metal is extracted from its ore, metal ions are changed to metal atoms.

Name this type of chemical reaction. **1**

(b) Iron can be extracted from its ore haematite, Fe_2O_3, in a blast furnace.

Calculate the percentage by mass of iron in haematite. **3**

Show your working clearly.

(c) Magnesium cannot be extracted from its ore in a blast furnace.

Suggest a method that would be suitable for the extraction of magnesium from its ore. **1**

Total marks **5**

MARKS | DO NOT WRITE IN THIS MARGIN

13. Sodium carbonate solution can be added to the water in swimming pools to neutralise the acidic effects of chlorine.

A student carried out a titration experiment to determine the concentration of a sodium carbonate solution.

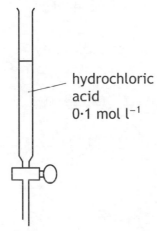

hydrochloric acid
0·1 mol l^{-1}

	Rough titre	1st titre	2nd titre
Initial burette reading (cm^3)	0·0	0·0	0·0
Final burette reading (cm^3)	16·5	15·9	16·1
Volume used (cm^3)	16·5	15·9	16·1

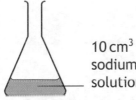

10 cm^3
sodium carbonate solution and indicator

(a) Using the results in the table, calculate the average volume, in cm^3, of hydrochloric acid required to neutralise the sodium carbonate solution.   **1**

(b) The equation for the reaction is

$$2HCl \quad + \quad Na_2CO_3 \quad \longrightarrow \quad 2NaCl \quad + \quad CO_2 \quad + \quad H_2O$$

Using your answer from part (a) calculate the concentration, in mol l^{-1}, of the sodium carbonate solution.   **3**

Show your working clearly.

Total marks   **4**

MARKS | DO NOT WRITE IN THIS MARGIN

14. Chemistry in the cinema.

In the film Dante's Peak, a family trapped by red hot lava escape by crossing a large lake in a boat made from aluminium. The volcano releases heat and the gases hydrogen chloride, sulfur dioxide and sulfur trioxide into the water in the lake. While crossing the lake, holes begin to appear in the bottom of the boat. Just after the family leave the boat, on the other side of the lake, the boat sinks.

Using your knowledge of chemistry, comment on whether or not the events described in the film could take place. 3

[END OF QUESTION PAPER]

MARKS | DO NOT WRITE IN THIS MARGIN

ADDITIONAL SPACE FOR ANSWERS AND ROUGH WORK

MARKS DO NOT WRITE IN THIS MARGIN

ADDITIONAL SPACE FOR ANSWERS AND ROUGH WORK

NATIONAL 5

2015

National Qualifications 2015

X713/75/02

Chemistry
Section 1—Questions

THURSDAY, 28 MAY

9:00 AM—11:00 AM

Instructions for the completion of Section 1 are given on *Page two* of your question and answer booklet X713/75/01.

Record your answers on the answer grid on *Page three* of your question and answer booklet.

Necessary data will be found in the Chemistry Data Booklet for National 5.

Before leaving the examination room you must give your question and answer booklet to the Invigilator; if you do not, you may lose all the marks for this paper.

SECTION 1

1. An atom has 26 protons, 26 electrons and 30 neutrons.

   The atom has

   A   atomic number 26, mass number 56

   B   atomic number 56, mass number 30

   C   atomic number 30, mass number 26

   D   atomic number 52, mass number 56.

2. The table shows the numbers of protons, electrons and neutrons in four particles, **W**, **X**, **Y** and **Z**.

Particle	Protons	Electrons	Neutrons
W	17	17	18
X	11	11	12
Y	17	17	20
Z	18	18	18

   Which pair of particles are isotopes?

   A   **W** and **X**

   B   **W** and **Y**

   C   **X** and **Y**

   D   **Y** and **Z**

3. Which of the following particles contains a different number of electrons from the others?

   You may wish to use the data booklet to help you.

   A Cl^-

   B S^{2-}

   C   Ar

   D Na^+

4. Which of the following diagrams shows the apparatus which would allow a soluble gas to be removed from a mixture of gases?

A

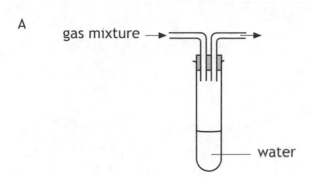

B

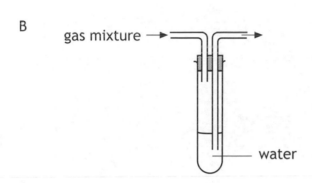

C

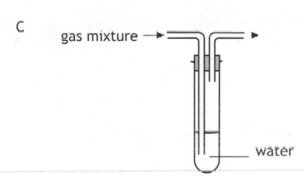

D

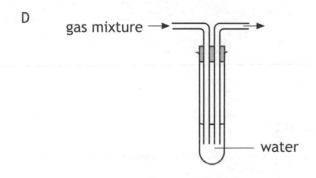

[Turn over

5. Which of the following diagrams could be used to represent the structure of a covalent network?

A

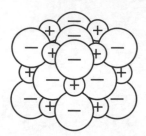

B

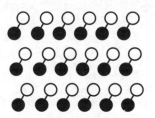

C

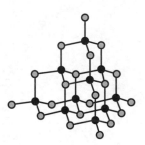

D

6. What is the charge on the chromium ion in $CrCl_3$?

A   1+

B   1−

C   3+

D   3−

7. The table contains information about calcium and calcium chloride.

	Melting point (°C)	Density (g cm^{-3})
Calcium	842	1·54
Calcium chloride	772	2·15

When molten calcium chloride is electrolysed at 800 °C the calcium appears as a

A   solid at the bottom of the molten calcium chloride

B   liquid at the bottom of the molten calcium chloride

C   solid on the surface of the molten calcium chloride

D   liquid on the surface of the molten calcium chloride.

8. $x\,Al(s)\ +\ y\,Br_2(\ell)\ \rightarrow\ z\,AlBr_3(s)$

This equation will be balanced when

A $x = 1$, $y = 2$ and $z = 1$

B $x = 2$, $y = 3$ and $z = 2$

C $x = 3$, $y = 2$ and $z = 3$

D $x = 4$, $y = 3$ and $z = 4$.

9. 0·2 mol of a gas has a mass of 12·8 g.

Which of the following could be the molecular formula for the gas?

A SO_2

B CO

C CO_2

D NH_3

[Turn over

10. Which of the following oxides, when shaken with water, would leave the pH unchanged? You may wish to use the data booklet to help you.

    A   Carbon dioxide

    B   Copper oxide

    C   Sodium oxide

    D   Sulfur dioxide

11. Which compound would **not** neutralise hydrochloric acid?

    A   Sodium carbonate

    B   Sodium chloride

    C   Sodium hydroxide

    D   Sodium oxide

12.

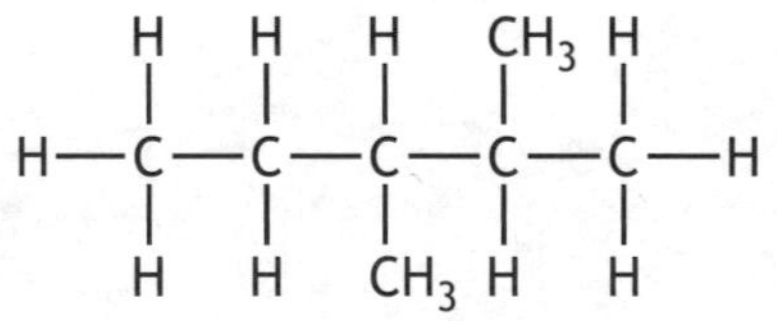

The name of the above compound is

    A   2,3-dimethylpropane

    B   3,4-dimethylpropane

    C   2,3-dimethylpentane

    D   3,4-dimethylpentane.

13. The shortened structural formula for an organic compound is

$$CH_3CH(CH_3)CH(OH)C(CH_3)_3$$

Which of the following is another way of representing this structure?

A

```
        H    H    OH   CH3
        |    |    |    |
   H — C  — C  — C  — C — CH3
        |    |    |    |
        H    CH3  H    CH3
```

B

```
        H    H    H    OH   CH3
        |    |    |    |    |
   H — C  — C  — C  — C  — C — CH3
        |    |    |    |    |
        H    H    H    H    CH3
```

C

```
        H    H    H    CH3  CH3
        |    |    |    |    |
   H — C  — C  — C  — C  — C — H
        |    |    |    |    |
        H    CH3  OH   H    H
```

D

```
        H    H    H    H    H    H
        |    |    |    |    |    |
   H — C  — C  — C  — C  — C  — C — CH3
        |    |    |    |    |    |
        H    CH3  OH   H    H    H
```

[Turn over

14. Three members of the cycloalkene homologous series are

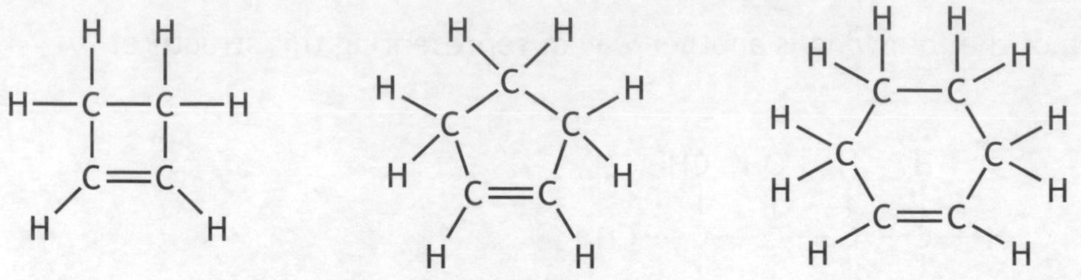

Which of the following is the general formula for this homologous series?

A C_nH_{2n-4}

B C_nH_{2n+2}

C C_nH_{2n}

D C_nH_{2n-2}

15. Metallic bonding is a force of attraction between

A   negative ions and positive ions

B   a shared pair of electrons and two nuclei

C   positive ions and delocalised electrons

D   negative ions and delocalised electrons.

16. Which pair of metals, when connected in a cell, would give the highest voltage and a flow of electrons from **X** to **Y**?

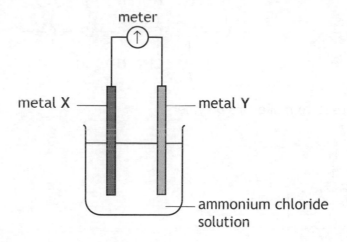

You may wish to use the data booklet to help you.

	Metal X	Metal Y
A	zinc	tin
B	tin	zinc
C	copper	magnesium
D	magnesium	copper

[Turn over

17. Part of the structure of a polymer is drawn below.

$$\begin{array}{c}
\quad\ \ H\quad\ H\quad\ H\quad\ H\quad\ H\quad\ H \\
\quad\ | \qquad | \qquad | \qquad | \qquad | \qquad | \\
-C-C-C-C-C-C- \\
\quad\ | \qquad | \qquad | \qquad | \qquad | \qquad | \\
\ CH_3\ \ H\ \ CH_3\ H\ \ CH_3\ H
\end{array}$$

The monomer used to make this polymer is

A
$$\begin{array}{c}
H\quad\ H \\
| \qquad\ | \\
C=C \\
| \qquad\ | \\
CH_3\ \ H
\end{array}$$

B
$$\begin{array}{c}
H\quad\ H \\
| \qquad\ | \\
-C-C- \\
| \qquad\ | \\
CH_3\ \ H
\end{array}$$

C
$$\begin{array}{c}
H\quad\ H\quad\ H \\
| \qquad\ | \qquad\ | \\
C=C-C-H \\
| \qquad\qquad\ | \\
CH_3\qquad\ CH_3
\end{array}$$

D
$$\begin{array}{c}
H\quad\ H\quad\ H \\
| \qquad\ | \qquad\ | \\
-C-C-C- \\
| \qquad\ | \qquad\ | \\
CH_3\ \ H\ \ CH_3
\end{array}$$

18. Sodium sulfate solution reacts with barium chloride solution.

$$Na_2SO_4(aq) \ + \ BaCl_2(aq) \longrightarrow BaSO_4(s) \ + \ 2NaCl(aq)$$

The spectator ions present in this reaction are

A Na^+ and Cl^-

B Na^+ and SO_4^{2-}

C Ba^{2+} and Cl^-

D Ba^{2+} and SO_4^{2-}.

19. Which of the following solutions would produce a precipitate when mixed together?
You may wish to use the data booklet to help you.

A   Ammonium chloride and potassium nitrate

B   Zinc nitrate and magnesium sulfate

C   Calcium nitrate and nickel chloride

D   Sodium iodide and silver nitrate

[Turn over for Question 20 on *Page twelve*

20. The table shows the colours of some ionic compounds in solution.

Compound	Colour
copper sulfate	blue
copper chromate	green
potassium chloride	colourless
potassium chromate	yellow

The colour of the chromate ion is

A   blue

B   green

C   colourless

D   yellow.

**[END OF SECTION 1. NOW ATTEMPT THE QUESTIONS IN SECTION 2
OF YOUR QUESTION AND ANSWER BOOKLET]**

N5

National
Qualifications
2015

Mark

X713/75/01

Chemistry
Section 1—Answer Grid
And Section 2

THURSDAY, 28 MAY

9:00 AM – 11:00 AM

Fill in these boxes and read what is printed below.

Full name of centre

Town

Forename(s)

Surname

Number of seat

Date of birth

Day	Month	Year	Scottish candidate number

Total marks — 80

SECTION 1 — 20 marks

Attempt ALL questions.

Instructions for the completion of Section 1 are given on *Page two*.

SECTION 2 —60 marks

Attempt ALL questions.

Necessary Data will be found in the Chemistry Data Booklet for National 5.

Write your answers clearly in the spaces provided in this booklet. Additional space for answers and rough work is provided at the end of this booklet. If you use this space you must clearly identify the question number you are attempting. Any rough work must be written in this booklet. You should score through your rough work when you have written your final copy.

Use **blue** or **black** ink.

Before leaving the examination room you must give this booklet to the Invigilator; if you do not, you may lose all the marks for this paper.

SQA

SECTION 1—20 marks

The questions for Section 1 are contained in the question paper X713/75/02.
Read these and record your answers on the answer grid on *Page three* opposite.
Use **blue** or **black** ink. Do NOT use gel pens or pencil.

1. The answer to each question is **either** A, B, C or D. Decide what your answer is, then fill in the appropriate bubble (see sample question below).

2. There is **only one correct** answer to each question.

3. Any rough work must be written in the additional space for answers and rough work at the end of this booklet.

Sample Question

To show that the ink in a ball-pen consists of a mixture of dyes, the method of separation would be

    A   fractional distillation

    B   chromatography

    C   fractional crystallisation

    D   filtration.

The correct answer is **B**—chromatography. The answer **B** bubble has been clearly filled in (see below).

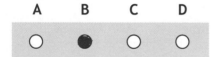

Changing an answer

If you decide to change your answer, cancel your first answer by putting a cross through it (see below) and fill in the answer you want. The answer below has been changed to **D**.

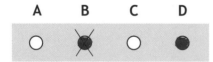

If you then decide to change back to an answer you have already scored out, put a tick (✓) to the **right** of the answer you want, as shown below:

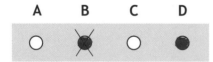

SECTION 1 — Answer Grid

	A	B	C	D
1	○	○	○	○
2	○	○	○	○
3	○	○	○	○
4	○	○	○	○
5	○	○	○	○
6	○	○	○	○
7	○	○	○	○
8	○	○	○	○
9	○	○	○	○
10	○	○	○	○
11	○	○	○	○
12	○	○	○	○
13	○	○	○	○
14	○	○	○	○
15	○	○	○	○
16	○	○	○	○
17	○	○	○	○
18	○	○	○	○
19	○	○	○	○
20	○	○	○	○

[Turn over

[BLANK PAGE]

DO NOT WRITE ON THIS PAGE

Page five

[Turn over for Question 1 on *Page six*

DO NOT WRITE ON THIS PAGE

MARKS | DO NOT WRITE IN THIS MARGIN

SECTION 2—60 marks

Attempt ALL questions

1. Ethyne is the first member of the alkyne family.

   It can be produced by the reaction of calcium carbide with water.

   The equation for this reaction is

   $$CaC_2(s) \ + \ 2H_2O(\ell) \ \longrightarrow \ C_2H_2(g) \ + \ Ca(OH)_2(aq)$$

   (a) The table shows the results obtained in an experiment carried out to measure the volume of ethyne gas produced.

Time (s)	0	30	60	90	120	150	180	210
Volume of ethyne (cm^3)	0	60	96	120	140	148	152	152

   Calculate the average rate of reaction between 60 and 90 seconds.

   Your answer must include the appropriate unit.

   Show your working clearly.

   3

MARKS | DO NOT WRITE IN THIS MARGIN

1. (continued)

(b) Draw a line graph of the results.

Use appropriate scales to fill most of the graph paper.

(Additional graph paper, if required, will be found on *Page twenty-seven.*)

3

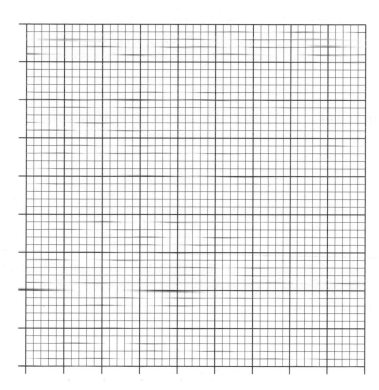

[Turn over

MARKS

DO NOT WRITE IN THIS MARGIN

2. Americium-241, a radioisotope used in smoke detectors, has a half-life of 432 years.

(a) The equation for the decay of americium-241 is

$$^{241}_{95}Am \longrightarrow {}^{4}_{2}He + X$$

Name element **X**. **1**

(b) Name the **type** of radiation emitted by the americium-241 radioisotope. **1**

(c) Another radioisotope of americium exists which has an atomic mass of 242.

Americium-242 has a half-life of 16 hours.

(i) A sample of americium-242 has a mass of 8 g.

Calculate the mass, in grams, of americium-242 that would be left after 48 hours. **2**

Show your working clearly.

(ii) Suggest why americium-241, and not americium-242, is the radioisotope used in smoke detectors. **1**

MARKS | DO NOT WRITE IN THIS MARGIN

3. Butter contains different triglyceride molecules.

   (a) A triglyceride molecule is made when the alcohol glycerol reacts with carboxylic acids.

      (i) Name the functional group present in glycerol.   1

      (ii) Name the family to which triglycerides belong.   1

   (b) When butter goes off, a triglyceride molecule is broken down, producing compounds X and Y.

$$HO-\overset{\overset{\displaystyle O}{\|}}{C}-C_3H_7 \qquad HO-\overset{\overset{\displaystyle O}{\|}}{C}-C_{17}H_{33}$$

X   Y

      (i) Name compound X.   1

      (ii) Describe the chemical test, including the result, to show that compound Y is unsaturated.   1

[Turn over

MARKS

4. Some sources of methane gas contain hydrogen sulfide, H_2S.

    (a) Draw a diagram, showing all outer electrons, to represent a molecule of hydrogen sulfide, H_2S.

    1

    (b) If hydrogen sulfide is not removed before methane gas is burned, sulfur dioxide is formed.

    When sulfur dioxide dissolves in water in the atmosphere, acid rain is produced.

    Circle the correct words to complete the sentence.

    Acid rain contains more $\left\{ \begin{array}{c} \text{hydrogen} \\ \text{hydroxide} \end{array} \right\}$ ions than $\left\{ \begin{array}{c} \text{hydrogen} \\ \text{hydroxide} \end{array} \right\}$ ions.

    1

    (c) In industry, calcium oxide is reacted with sulfur dioxide to reduce the volume of sulfur dioxide released into the atmosphere.

    Explain why calcium oxide is able to reduce the volume of sulfur dioxide gas released.

    2

MARKS

5. A researcher investigated the conditions for producing ammonia.

$$N_2(g) + 3H_2(g) \rightleftharpoons 2NH_3(g)$$

(a) Name the catalyst used in the production of ammonia.

1

(b) In her first experiment she measured how the percentage yield of ammonia varied with pressure at a constant temperature of 500 °C.

Pressure (atmospheres)	100	200	300	400	500
Percentage yield (%)	10	18	26	3?	40

Predict the percentage yield of ammonia at 700 atmospheres.

1

(c) In a second experiment the researcher kept the pressure constant, at 200 atmospheres, and changed the temperature as shown.

Temperature (°C)	200	300	400	500
Percentage yield (%)	89	67	39	18

Describe how the percentage yield varies with temperature.

1

(d) **Using the information in both tables**, suggest the combination of temperature and pressure that would produce the highest percentage yield of ammonia.

1

[Turn over

MARKS | DO NOT WRITE IN THIS MARGIN

6. Read the passage below and answer the questions that follow.

Clean coal technology comes a step closer

It is claimed a process called Coal-Direct Chemical Looping (CDCL) is able to release energy from coal while capturing 99% of the carbon dioxide emitted. CDCL works by extracting the energy from coal using a reaction other than combustion.

A mixture of powdered coal and beads of iron(III) oxide is heated inside a metal cylinder. Carbon in the coal and oxygen from the beads react to form carbon dioxide which can be captured for recycling or stored.

This reaction gives off heat energy that could be used to heat water in order to drive electricity-producing steam turbines.

Adapted from Focus: Science and Technology, April 2013

(a) The CDCL process produced 300 tonnes of carbon dioxide.

Calculate the mass, in tonnes, of carbon dioxide released into the atmosphere.   **1**

(b) Write the ionic formula for the iron compound used in CDCL.   **1**

(c) State the term used to describe all chemical reactions that release heat energy.   **1**

MARKS | DO NOT WRITE IN THIS MARGIN

7. A student was asked to carry out an experiment to determine the concentration of a copper(II) sulfate solution.

Part of the work card used is shown.

Determination of the Concentration of Copper(II) Sulfate Solution

1.   Weigh an empty crucible

2.   Add 100 cm^3 copper(II) sulfate solution

3.   Evaporate the solution to dryness

4.   Weigh the crucible containing dry copper(II) sulfate

(a) Suggest how the student could have evaporated the solution to dryness.   **1**

(b) The student found that the 100 cm^3 solution contained 3·19 g of copper(II) sulfate, $CuSO_4$.

Calculate the concentration of the solution in mol l^{-1}.   **2**

Show your working clearly.

[Turn over

MARKS | DO NOT WRITE IN THIS MARGIN

8. A student calculated the energy absorbed by water when ethanol is burned using two different methods.

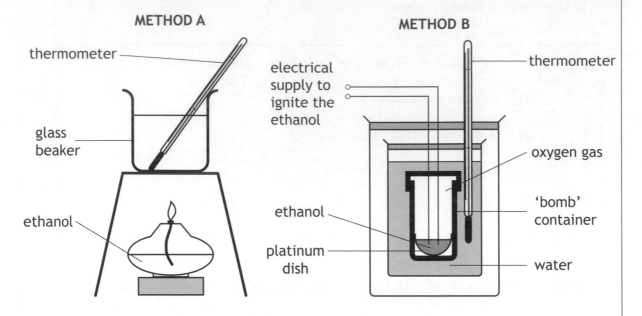

METHOD A

thermometer

glass beaker

ethanol

electrical supply to ignite the ethanol

METHOD B

thermometer

oxygen gas

ethanol

'bomb' container

platinum dish

water

The student recorded the following data.

	Method	
	A	B
Mass of ethanol burned (g)	0·5	0·5
Mass of water heated (g)	100	100
Initial temperature of water (°C)	24	24
Final temperature of water (°C)	32	58

(a) The final temperature of water in method **B** is higher than in method **A**.

Suggest why there is a difference in the energy absorbed by the water. **1**

MARKS | DO NOT WRITE IN THIS MARGIN

8.   (continued)

(b) Calculate the energy, in kJ, absorbed by the water in method **B**.

You may wish to use the data booklet to help you.

Show your working clearly.

3

[Turn over

MARKS | DO NOT WRITE IN THIS MARGIN

9. Aluminium can be extracted from naturally occurring metal compounds such as bauxite.

(a) State the term used to describe naturally occurring metal compounds such as bauxite.   1

(b) Bauxite is refined to produce aluminium oxide.

Electrolysis of molten aluminium oxide produces aluminium and oxygen gas.

The ion-electron equations taking place during the electrolysis of aluminium oxide are

$$Al^{3+} + 3e^- \longrightarrow Al$$

$$2O^{2-} \longrightarrow O_2 + 4e^-$$

(i) Write the redox equation for the overall reaction.   1

(ii) State why ionic compounds, like aluminium oxide, conduct electricity when molten.   1

MARKS | DO NOT WRITE IN THIS MARGIN

9. (continued)

(c) Bauxite contains impurities such as silicon dioxide.

Silicon can be extracted from silicon dioxide as shown.

$$SiO_2 \quad + \quad 2Mg \quad \longrightarrow \quad Si \quad + \quad 2MgO$$

Identify the reducing agent in this reaction.

1

[Turn over

MARKS | DO NOT WRITE IN THIS MARGIN

10. A group of students were given strips of aluminium, iron, tin and zinc.

Using your knowledge of chemistry, suggest how the students could identify each of the four metals.

3

MARKS | DO NOT WRITE IN THIS MARGIN

11. Electrons can be removed from all atoms

The energy required to do this is called the ionisation energy.

The first ionisation energy for an element is defined as the energy required to remove one mole of electrons from one mole of atoms, in the gaseous state.

The equation for the first ionisation energy of chlorine is

$$Cl(g) \longrightarrow Cl^+(g) + e^-$$

(a) State the electron arrangement for the Cl^+ ion.   **1**

You may wish to use the data booklet to help you.

(b) Write the equation for the first ionisation energy of magnesium.   **1**

(c) Information on the first ionisation energy of some elements is given in the table.

Element	First ionisation energy (kJ mol^{-1})
lithium	526
fluorine	1690
sodium	502
chlorine	1260
potassium	425
bromine	1150

Describe the trend in the first ionisation energy going down a group in the Periodic Table.   **1**

[Turn over

MARKS | DO NOT WRITE IN THIS MARGIN

12. The structural formulae of two hydrocarbons are shown.

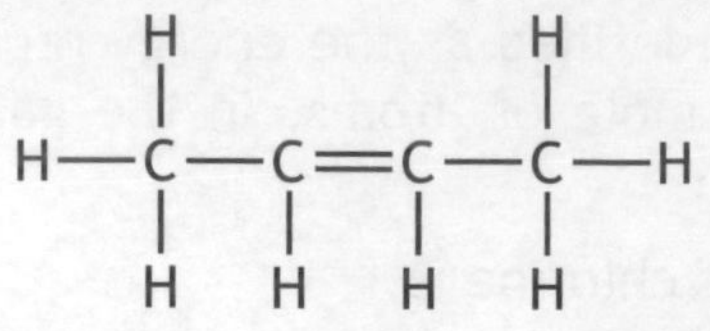

A 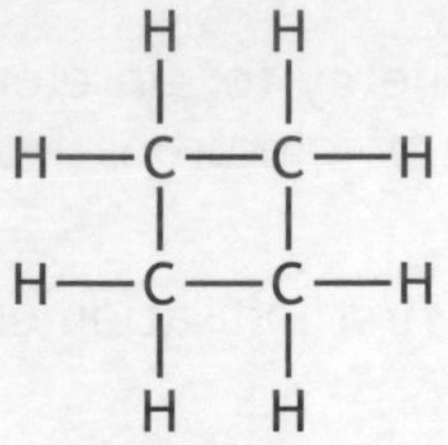   B

(a) Name hydrocarbon **A**.

1

(b) Hydrocarbons **A** and **B** can be described as isomers.

State what is meant by the term isomer.

1

MARKS | DO NOT WRITE IN THIS MARGIN

12. (continued)

(c) Hydrocarbon **A** can undergo an addition reaction with water to form butan-2-ol as shown.

$$
\begin{array}{ccc}
& \quad H \qquad\quad H & \\
& \quad | \qquad\qquad | & \\
H-C-C=C-C-H & \quad + & \quad H-OH \\
\;\; |\quad\; |\quad\; |\quad\; | & & \\
\;\; H\;\; H\;\; H\;\; H & &
\end{array}
$$

↓

$$
\begin{array}{c}
H \quad H \;\; OH \; H \\
| \quad\; | \quad\; | \quad\; | \\
H-C-C-C-C-H \\
| \quad\; | \quad\; | \quad\; | \\
H \quad H \quad H \quad H
\end{array}
$$

A similar reaction can be used to produce 3-methylpentan-3-ol.

Draw a structural formula for the hydrocarbon used to form this molecule.   **1**

[dashed answer box]   +   H—OH
water

↓

$$
\begin{array}{c}
H \quad H \;\; OH \; H \quad H \\
| \quad\; | \quad\; | \quad\; | \quad\; | \\
H-C-C-C-C-C-H \\
| \quad\; | \quad\; | \quad\; | \quad\; | \\
H \quad H \quad | \quad H \quad H \\
\qquad\;\; H-C-H \\
\qquad\qquad | \\
\qquad\qquad H
\end{array}
$$

3-methylpentan-3-ol

[Turn over

MARKS | DO NOT WRITE IN THIS MARGIN

13. Succinic acid is a natural antibiotic.

The structure of succinic acid is shown.

$$HO-\overset{\overset{\displaystyle O}{\|}}{C}-\overset{\overset{\displaystyle H}{|}}{\underset{\underset{\displaystyle H}{|}}{C}}-\overset{\overset{\displaystyle H}{|}}{\underset{\underset{\displaystyle H}{|}}{C}}-\overset{\overset{\displaystyle O}{\|}}{C}-OH$$

(a) Name the functional group present in succinic acid.   1

(b) Succinic acid can form a polymer with ethane-1,2-diol.

The structure of ethane-1,2-diol is shown.

$$H-O-\overset{\overset{\displaystyle H}{|}}{\underset{\underset{\displaystyle H}{|}}{C}}-\overset{\overset{\displaystyle H}{|}}{\underset{\underset{\displaystyle H}{|}}{C}}-O-H$$

(i) Name the type of polymerisation which would take place between succinic acid and ethane-1,2-diol.   1

(ii) Draw the repeating unit of the polymer formed between succinic acid and ethane-1,2-diol.   1

MARKS | DO NOT WRITE IN THIS MARGIN

14. Titanium is the tenth most commonly occurring element in the Earth's crust.

(a) The first step in the extraction of titanium from impure titanium oxide involves the conversion of titanium oxide into titanium(IV) chloride.

$$TiO_2 + 2Cl_2 + 2C \longrightarrow TiCl_4 + 2X$$

(i) Identify X.   1

(ii) Titanium(IV) chloride is a liquid at room temperature and does not conduct electricity.

Suggest the type of bonding that is present in titanium(IV) chloride.   1

(b) The next step involves separating pure titanium(IV) chloride from other liquid impurities that are also produced during the first step.

Suggest a name for this process.   1

(c) The equation for the final step in the extraction of titanium is

$$TiCl_4 + 4Na \longrightarrow Ti + 4NaCl$$

The sodium chloride produced can be electrolysed.

Suggest how this could make the extraction of titanium from titanium oxide more economical.   1

[Turn over

15. Vitamin C is found in fruits and vegetables.

Using iodine solution, a student carried out titrations to determine the concentration of vitamin C in orange juice.

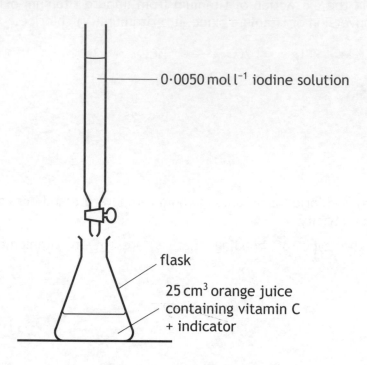

0·0050 mol l⁻¹ iodine solution

flask

25 cm³ orange juice containing vitamin C + indicator

The results of the titration are given in the table.

Titration	Initial burette reading (cm³)	Final burette reading (cm³)	Titre (cm³)
1	1·2	18·0	16·8
2	18·0	33·9	15·9
3	0·5	16·6	16·1

(a) Calculate the average volume, in cm³, that should be used in calculating the concentration of vitamin C.

1

MARKS | DO NOT WRITE IN THIS MARGIN

15. **(continued)**

(b) The equation for the reaction is

$$C_6H_8O_6(aq) \quad + \quad I_2(aq) \quad \longrightarrow \quad C_6H_6O_6(aq) \quad + \quad 2HI(aq)$$

vitamin C

Calculate the concentration, in $mol\, l^{-1}$, of vitamin C in the orange juice.   **3**
Show your working clearly.

[Turn over for Question 16 on *Page twenty-six*

MARKS

DO NOT WRITE IN THIS MARGIN

16. A student is given three different compounds each containing carbon.

Using your knowledge of chemistry, describe how the student could identify the compounds.

3

[END OF QUESTION PAPER]

MARKS | DO NOT WRITE IN THIS MARGIN

ADDITIONAL SPACE FOR ANSWERS

Question 1(b)

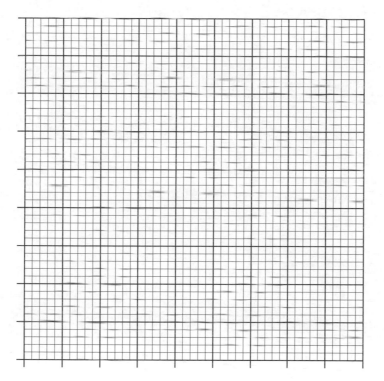

MARKS

ADDITIONAL SPACE FOR ANSWERS AND ROUGH WORK

MARKS | DO NOT WRITE IN THIS MARGIN

ADDITIONAL SPACE FOR ANSWERS AND ROUGH WORK

[BLANK PAGE]

DO NOT WRITE ON THIS PAGE

[BLANK PAGE]

DO NOT WRITE ON THIS PAGE

[BLANK PAGE]

DO NOT WRITE ON THIS PAGE

Section 1

Question	Response
1.	D
2.	A
3.	A
4.	D
5.	D
6.	A
7.	C
8.	C
9.	B
10.	C
11.	C
12.	A
13.	A
14.	C
15.	A
16.	C
17.	D
18.	D
19.	B
20.	B

Section 2

1. (a) (i) $\dfrac{32 - 12}{8}$

   = 2.5 litres per microsecond

   2

   (ii) 4.0 (±1) microseconds

   1

   (b) $NaN_3 \rightarrow Na + N_2$

   1

   (c) Explosive/
   Highly reactive/very reactive
   or
   Flammable

   1

2. The maximum available mark would be awarded to a student who has demonstrated a good understanding of the chemistry involved. The student shows good comprehension of the chemistry of the situation and has provided a logically correct answer to the question posed.

   3

3. (a) A workable diagram:
   • Syringe (must have plunger)
   • or Displacement of water into a vertical measuring cylinder/graduated test tube
   • Diagram must not have closed-off tubes

   1

   (b) Calcium chloride

   1

(c)

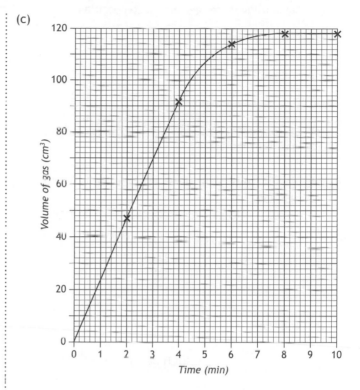

1 mark for correct axes labels and units
1 mark for scale on X and Y axis
1 mark for graph drawn accurately

3

4. (a) 11 – proton
   13 – neutron

   1

   (b) To achieve a stable electron arrangement/full outer energy level/noble gas arrangement

   1

   (c) (i)

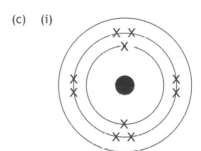

   1

   (ii) • The attraction/pull/electrostatic force to the positively charged nucleus (and the (negatively charged) electrons)
   • Attraction/pull/electrostatic force between (positive) protons and electrons

   1

5. (a) (i) 25–30

   1

   (ii) Branched hydrocarbons burn more efficiently or the greater the number of carbons the lower the octane number efficiency

   1

   (b) $Eh = cm\Delta T = 4.18 \times 1 \times 58 = 242.44$ kJ

   3

6. (a) Fertilisers   1

   (b) Hypoxic Dead Zone   1

   (c) Algae using up oxygen and decomposition   1

   (d) Three times as high/higher   1

7. (a) metal 3 circled   1

   (b) The reading would be 0 V   1

   (c) Glucose does not conduct or glucose is covalent. No
       ions, only molecules   1

8. (a) Diagram must show three monomer units linked
       together

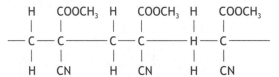

                                                                               1

   (b) Addition   1

   (c)
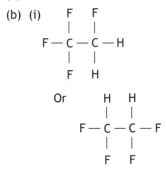

                                                                               1

9. (a) Tetrahedral/Tetrahedron

   (b) (i)
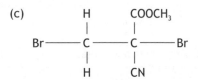

       Or
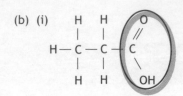

       Or CF_3CH_2F or CHF_2CHF_2 1

       (ii) Chlorine/Cl/Cl_2 1

       (iii) Shorter atmospheric life/breaks down faster   1

10. (a) 1.62g on its own = 3 marks

        2 moles Al → 6 moles Ag

        54g → 648g

        1g → 648 ÷ 54g

        0.135g → 648 ÷ 54 × 0.135g

        = 1.62g   3

    (b) Weigh mass of beaker at start and again at the end.
        (Should have decreased.)
        Find mass difference   1

11. (a) Addition / additional   1

    (b) (i)
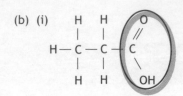
                                                                               1

        (ii) Carboxylic acid   1

    (c)
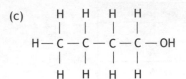

        Or
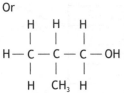

        Or

        H   H   H
        | | |
    H — C — C — C — OH
        | | |
        H CH_3 H

                                                                               1

12. (a) Same atomic number, different mass number   1

    (b) $^{238}_{92}U \rightarrow ^{234}_{90}Th + ^{4}_{2}He$ 2

    (c) Time take for the activity/ quantity of a radioactive
        sample to fall by half.   1

13. The maximum available mark would be awarded to a
    student who has demonstrated a good understanding
    of the chemistry involved. The student shows good
    comprehension of the chemistry of the situation and
    has provided a logically correct answer to the question
    posed.   3

14. (a) $\dfrac{25.1 + 24.9}{2}$

        25.0 cm³ on its own 1 mark   1

    (b) 3 marks 0.25 mol l⁻¹ on its own
        1 mole of HCl → 1 mole of NaOH
        0.0025 moles of HCl → 0.0025 moles of NaOH
        Concentration of NaOH = 0.0025/ 0.01 = 0.25 mol l⁻¹   3

15. (a) $2KOH + H_2SO_4 \rightarrow K_2SO_4 + 2H_2O$ 1

    (b) Neutralisation   1

    (c) 3 marks on its own 44.8%
        1 mole of K_2SO_4 = 174

        $\dfrac{78}{174} \times 100 = 44.8\%$ 3

NATIONAL 5 CHEMISTRY MODEL PAPER 2

Section 1

Question	Response
1.	B
2.	A
3.	B
4.	D
5.	B
6.	A
7.	A
8.	C
9.	A
10.	C
11.	A
12.	C
13.	B
14.	B
15.	C
16.	D
17.	A
18.	D
19.	B
20.	C

Section 2

1. (a) Covalent

   1

   (b) (i) $TiCl_4 + 4Na \rightarrow Ti + 4NaCl$

   1

   (ii) Sodium is more reactive than titanium

   1

2. (a) (i) 2+

   1

   (ii) 3 marks on its own 86.6%
   I mole of PbS – 239

   $\frac{207}{239} \times 100 = 86.6\%$

   3

   (b)

Metal	Method of extraction
mercury	using heat alone
aluminium	electrolysis of molten ore
copper	heating with carbon

   1

3. (a) 2 marks for on its own 8

   $\frac{6.72}{0.2 \times 4.2} = 8°C$

   2

   (b) $NH_4^+ NO_3^-$

   1

4. (a) Covalent bonding/covalent network

   2

   (b)

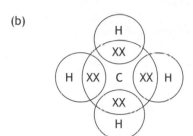

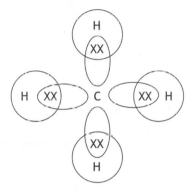

   1

5. (a) Family of hydrocarbons with similar chemical properties and same general formula

   1

   (b) 3 marks on its own – 37·4 kJ
   Eh = cmΔT = 4·18 × 0·2 × 44·7 = 37·4 kJ

   3

6. The maximum available mark would be awarded to a student who has demonstrated a good understanding of the chemistry involved. The student shows good comprehension of the chemistry of the situation and has provided a logically correct answer to the question posed.

   3

7. (a) (i) All three required:
   - LHS = copper/Cu
   - Top RHS = Iron/Fe
   - Bottom RHS = 100 cm³ 0·1 mol l⁻¹

   1

   (ii) Repeated to allow averages/mean to be calculated

   1

   (b) (i) From right to left → arrow should be on wires or very close to it

   1

   (ii) Reduction

   1

8. (a) Exothermic

   1

   (b) (i) 1 mark for correct axes labels and units
   1 mark for scale on X and Y axis
   1 mark for graph drawn accurately

   3

   (ii) 13g or correct reading from graph drawn

   1

   (c) Method of reducing heat loss from beaker to the surroundings/stirring or other suitable answer.

   1

9. (a) $^3_1H \rightarrow \ ^3_2He + \ ^0_{-1}e$

1

(b) 2 marks for 24·6 years on its own
2 half-lives

2

10. (a)

```
      H  OH  H
      |   |   |
  H — C — C — C — H
      |   |   |
      H   H   H
```

1

(b) Propan-1-ol

1

(c) Hydroxyl

1

(d) 0·01 moles on its own 2 marks
0·46/46 = 0·01

2

(e) Carboxylic acids/alkanoic acids

1

11. (a) Protection from UV/removal of contaminants/self-cleaning of the atmosphere

1

(b) $Br_2 + O_3 \rightarrow Br_2O + O_2$

1

(c) 3 marks 96 g on its own
1 mole of Cl → 1 mole of O_3
2 moles of Cl → 2 moles of O_3
mass of O_3 = 48 x 2 = 96 g

3

12. (a)

```
      H
      |     O
      |    //
  H — C — C
      |    \
      H     OH
```

1

(b) 3 marks 1 mol l⁻¹ on its own
1 mole of NaOH→ 1 mole of CH_3COOH
0·01 moles of NaOH→ 0·01 moles of CH_3COOH
Concentration of CH_3COOH = 0·01/ 0·01 = 1 mol l⁻¹

3

(c) The maximum available mark would be awarded to a student who has demonstrated a good understanding of the chemistry involved. The student shows good comprehension of the chemistry of the situation and has provided a logically correct answer to the question posed.

3

13. (a) (i)

```
   H          H  O          O
   |          |  ||         ||
 — N —⬡— N — C —⬡— C —
```

1

(ii) Condensation

1

(b)

```
        H  H
        |  |          O          O
        |  |          ||         ||
 H—O—C—C—O—H    C          C
        |  |         /  \___/  \
        H  H       HO          OH
```

2

14. (a) 46 g

1

(b) (i) 46 − 37 = 9 g

2

(ii) Filtration/filter/filtering

1

Section 1

Question	Response
1.	A
2.	B
3.	D
4.	B
5.	B
6.	B
7.	C
8.	B
9.	D
10.	A
11.	A
12.	D
13.	A
14.	D
15.	A
16.	D
17.	D
18.	A
19.	B
20.	C

Section 2

1. (a) $Ag^+(aq) + e^- \rightarrow Ag \ (s)$

1

(b)

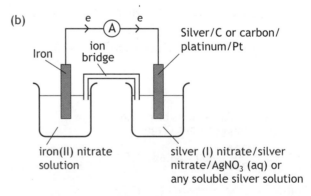

Silver/C or carbon/platinum/Pt

silver (I) nitrate/silver nitrate/AgNO₃ (aq) or any soluble silver solution

1

(c) To complete the circuit/allow ions to flow

1

2. (a) $2Mg + UF_4 \rightarrow 2MgF_2 + U$

1

(b) Magnesium would burn/react with oxygen

1

(c) Covalent

1

(d) The maximum available mark would be awarded to a student who has demonstrated a good understanding of the chemistry involved. The student shows good comprehension of the chemistry of the situation and has provided a logically correct answer to the question posed.

3

3. (a) $CuCO_3 + H_2SO_4 \rightarrow CuSO_4 + CO_2 + H_2O$

1

(b)

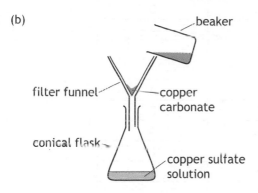

1

(c) pH neutral, no more gas produced/no more fizzing

1

(d) To ensure that all the acid has reacted.

1

4. (a) Hydroxyl

1

(b)

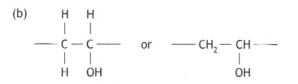

1

(c) Addition

1

5. (a) Potassium permanganate
Water
(Conc.) Sulfuric acid

1

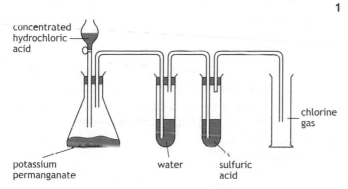

(b) (i) As the atomic number increases the melting point increases

1

(ii) $470°C \pm 20$

1

6. (a) Ethyne

1

(b) (i)

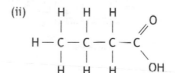

1

(ii) Bromines are not attached to adjacent carbon atoms

1

7. (a) (i) 25(%)

1

(ii) $25/100 \times 6 = 1\cdot5$ g

1

(b) $1\cdot24$ g on its own 3 marks
4 moles to 2 moles
no of moles of Ag $= 1\cdot08/108$
$= 0\cdot01$ moles
no of moles of $Ag_2S = 0\cdot01/2$
$= 0\cdot005$
GFM Ag_2S $= 248$
Mass of Ag_2S $= 0\cdot005 \times 248$
$= 1\cdot24$ g

3

8. (a)

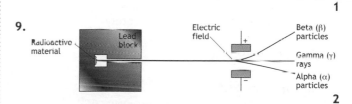

1

(b) Heptan-4-one

1

(c) 140–160

1

(d) (i) Solvents or perfumes or materials

1

(ii)

$$H-\underset{\underset{H}{|}}{\overset{\overset{H}{|}}{C}}-\underset{\underset{H}{|}}{\overset{\overset{H}{|}}{C}}-\underset{\underset{H}{|}}{\overset{\overset{H}{|}}{C}}-C\overset{\displaystyle O}{\underset{\displaystyle OH}{\big<}}$$

1

9.

2

10. (a) Hydrogen, nitrogen, argon and helium

1

(b) Doesn't form bonds/stable electron arrangement

1

(c) Hundred thousand times faster

1

(d) propan-1-ol and propan-2-ol

2

11. (a) Network or Lattice   1

    (b) Sb_2O_3
        $(Sb^{3+})_2(O^{2-})_3$ 1

    (c) (i) $^{11}_{5}B$ 1

        (ii) Isotopes   1

12. (a) Haber   1

    (b) Water/H_2O 1

    (c) 35% on its own 3 marks
        $\% = \dfrac{28 \times 100}{80} = 35$ 3

13. The maximum available mark would be awarded to a
    student who has demonstrated a good understanding
    of the chemistry involved. The student shows good
    comprehension of the chemistry of the situation and
    has provided a logically correct answer to the question
    posed.   3

14. (a)

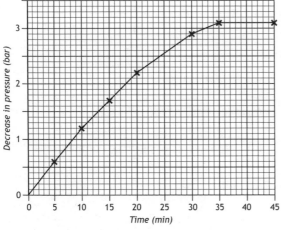

    1 mark for correct axes labels and units
    1 mark for scale on X and Y axis
    1 mark for graph drawn accurately   3

    (b) (i) 35 seconds   1

        (ii) $\dfrac{2 \cdot 2 - 1 \cdot 2}{10}$
             = 0.1 bar min^{-1} 2

15. $Eh = cm\Delta T = 4 \cdot 18 \times 0 \cdot 1 \times 7 = 2 \cdot 93$ kJ   3

NATIONAL 5 CHEMISTRY 2014

Section 1

Question	Response
1.	A
2.	D
3.	A
4.	C
5.	C
6.	A
7.	D
8.	A
9.	C
10.	B
11.	C
12.	B
13.	C
14.	C
15.	A
16.	D
17.	B
18.	B
19.	B
20.	D

Section 2

1. (a) **Repulsion/repelled** by nucleus/positive nucleus
       /protons/positive protons/positive particles in
       nucleus or in atom or in gold/ like charges in
       nucleus, atom or gold

   (b) (i) Protons – 79
           Electrons – 79
           Neutrons – 118

       (ii) Same atomic number/protons AND different
            mass number/mass/number of neutrons

            Atoms of the same element with different mass
            number/mass/number of neutrons

2. (a) Covalent network
       Ionic lattice
       Metallic lattice
       (Discrete) covalent molecular

   (b) Delocalised /able or free to move

3. (a) Potassium is an essential element
       or
       humans/human body cannot store it/have no
       mechanism for storing it

   (b) 0·022 or 0·02

   (c) Lilac/purple

   (d) $K^+ NO_3^-$

4. (a)

$NC_{12}H_8$	H	$NC_{12}H_8$	H	$NC_{12}H_8$	H
|	|	|	|	|	|
—C———	C—C———	C—C———	C—

with H atoms below each carbon

(b) Addition

5. (a) alpha or α

(b) ¼/0·25/25%

(c) Sodium/Na

$^{24}_{11}Na$ $\quad ^{24}Na \quad _{11}Na$

6. This is an open ended question

1 mark: The student has demonstrated a limited understanding of the chemistry involved. The candidate has made some statement(s) which is/are relevant to the situation, showing that at least a little of the chemistry within the problem is understood.

2 marks: The student has demonstrated a reasonable understanding of the chemistry involved. The student makes some statement(s) which is/are relevant to the situation, showing that the problem is understood.

3 marks: The maximum available mark would be awarded to a student who has demonstrated a good understanding of the chemistry involved. The student shows a good comprehension of the chemistry of the situation and has provided a logically correct answer to the question posed. This type of response might include a statement of the principles involved, a relationship or an equation, and the application of these to respond to the problem. This does not mean the answer has to be what might be termed an "excellent" answer or a "complete" one.

7 (a) (i) Haber

(ii) Diagram showing three hydrogen atoms and one nitrogen atom with three pairs of bonding electrons and two non-bonding electrons in nitrogen eg

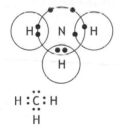

H :C̈: H
     H

(b) (i) Water/H_2O/Hydrogen oxide

(ii) Arrow from nitrogen monoxide from absorber to nitrogen monoxide below reactor (anywhere below the reactor and above nitrogen dioxide)

(c) (i) Neutralisation

(ii) Evaporation

or

boil it/boil off the water

or

distillation

8. (a) Perfumes, solvents, flavourings, fragrances, preservatives

(b) (i) Hydroxyl

(ii) Any correct full or shortened structural formula for an isomer

(iii) $C_nH_{2n}O_2$

$C_nH_{2n+1}COOH$

(c) ethanol

propanoic acid

9. (a) They have similar chemical properties

and

They have the same general formula.

(b) Butane, or it, has stronger/more/bigger forces of attraction between molecules or mention of intermolecular attractions

(c) 2090

(d) Produces SO_2/acidic gases/oxides of sulfur

Produces acid rain

10. (a) (i) The higher/lower the number of carbon atoms the higher/lower the flash point

The flash point increases/decreases as the number of carbon atoms increases/decreases

(ii) 47 – 51 inclusive

(b) 99

11. (a) $2Cl^- \longrightarrow Cl_2 + 2e^-$

or

$2Cl^- - 2e^- \longrightarrow Cl_2$

(b) (i) sodium hydroxide

or

sodium oxide

(ii) Water is the only product

Hydrogen is infinite/renewable

Doesn't produce greenhouse gases/CO_2/CO

The products/gases produced do not contribute to the greenhouse effect/global warming

Fossil fuels not being used up as fuel

(c)

12. (a) Reduction

(b) 70

(c) Electrolysis

13. (a) 16

(b) 0·08

14. This is an open ended question

1 mark: The student has demonstrated a limited understanding of the chemistry involved. The candidate has made some statement(s) which is/are relevant to the situation, showing that at least a little of the chemistry within the problem is understood.

2 marks: The student has demonstrated a reasonable understanding of the chemistry involved. The student makes some statement(s) which is/are relevant to the situation, showing that the problem is understood.

3 marks: The maximum available mark would be awarded to a student who has demonstrated a good understanding of the chemistry involved. The student shows a good comprehension of the chemistry of the situation and has provided a logically correct answer to the question posed. This type of response might include a statement of the principles involved, a relationship or an equation, and the application of these to respond to the problem. This does not mean the answer has to be what might be termed an "excellent" answer or a "complete" one.

NATIONAL 5 CHEMISTRY 2015

Section 1

Question	Response
1.	A
2.	B
3.	D
4.	C
5.	D
6.	C
7.	C
8.	B
9.	A
10.	B
11.	B
12.	C
13.	A
14.	D
15.	C
16.	D
17.	A
18.	A
19.	D
20.	D

Section 2

1. (a) $0.8 \text{ cm}^3 \text{ s}^{-1}$ **or** $0.8 \text{ cm}^3/\text{s}$

(b) Both axes labelled with units   1
Both scales   1
Graph drawn accurately   1
The line must be drawn from the origin.

2. (a) Neptunium **or** Np
or
^{237}Np $^{237}_{93}\text{Np}$ $_{93}\text{Np}$

(b) Alpha **or** α **or** $^{4}_{2}\alpha$

(c) (i) 1

(ii) (It/Americium 241/Am-241) has a long/longer half life
or
will not need to be replaced as often **or** words to this effect
or
(It/Americium 241/Am-241) emits alpha radiation (particles) which has a low penetrating power/ doesn't travel far/stopped by the smoke particles.

3. (a) (i) Hydroxyl **or** OH **or** –OH

(ii) Ester **or** esters **or** fats **or** oils

(b) (i) Butanoic acid
   or
   methylpropanoic acid
   or
   2-methylpropanoic acid
   or
   butyric acid

   (ii) Bromine/Br_2 decolourised/discolourised
   or
   bromine/Br_2 goes colourless

4. (a) Diagram showing two hydrogen atoms and one sulfur atom with two pairs of bonding electrons and two non-bonding pair of electrons in sulfur e.g.

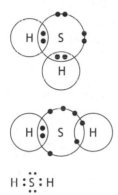

$$H : \overset{\cdot\cdot}{\underset{\cdot\cdot}{S}} : H$$

(b) 1st = hydrogen
   2nd = hydroxide

   Both required for 1 mark

(c) It/calcium oxide is a base
   or
   forms an alkaline solution (alkali) when dissolved in water. For the mention of alkali the candidate must explicitly state the calcium oxide is in solution/dissolved in water   1

   and
   Mention of it neutralising sulfur dioxide/it neutralises it/
   or
   a neutralisation reaction takes place.   1

5. (a) Iron **or** Fe

(b) Any value from 52–56 inclusive

(c) As temperature increases the yield decreases.
   or
   As temperature decreases the yield increases.
   or
   The yield increases as the temperature decreases.
   or
   The yield decreases as the temperature increases.
   Accept percentage in place of yield.

(d) temperature 200 °C **or** a value **below** 200 °C
   and
   pressure 500 atmospheres **or** a value **greater** than 500 atmospheres

   Both required for 1 mark

6. (a) 3

(b) $(Fe^{3+})_2(O^{2-})_3$
   or
   $Fe^{3+}_2\ O^{2-}_3$
   or
   $(Fe^{3+})_2\ O^{2-}_3$

   or
   $Fe^{3+}_2(O^{2-})_3$
   or
   $Fe_2^{3+}\ O_3^{2-}$

(c) Exothermic **or** exothermal

7. (a) Boil it
   or boil off the water
   or heat it
   or leave it for some time/overnight/next lesson
   or leave it on the window ledge
   or use Bunsen (burner)
   or appropriate diagram

(b) 0·2

8. (a) Method B (it)
   Complete combustion/more oxygen/pure oxygen
   Less/no heat loss (to surroundings)
   Better insulation
   Metal/platinum is a better conductor
   or
   Method A
   Incomplete combustion
   Less oxygen
   (More) heat loss to surroundings
   No draught shield/no insulation
   Glass is a poor conductor
   Flame too far away from beaker
   or
   Any other reasonable answer

(b) 11 / 14·7 / 14·71 / 11·212

9. (a) (Metal) ore/ores

(b) (i) $4Al^{3+} + 6O^{2-} \rightarrow 4Al + 3O_2$
   (or correct multiples)

   All must be correct for 1 mark

   (ii) Ions free to move
   or
   ions able to move
   or
   ions mobile

(c) Mg
   or
   magnesium
   or
   2Mg
   or
   Mg circled/highlighted/underlined in equation.

10. This is an open ended question
   1 mark: The student has demonstrated a limited understanding of the chemistry involved. The candidate has made some statement(s) which is/are relevant to the situation, showing that at least a little of the chemistry within the problem is understood.
   2 marks: The student has demonstrated a reasonable understanding of the chemistry involved. The student makes some statement(s) which is/are relevant to the situation, showing that the problem is understood.
   3 marks: The maximum available mark would be awarded to a student who has demonstrated a good understanding of the chemistry involved. The student shows a good comprehension of the chemistry of the situation and has provided a logically correct answer to the question posed. This type of response might include a statement of the principles involved, a relationship or an equation, and the application of these to respond to the problem. This does not mean the answer has to be what might be termed an "excellent" answer or a "complete" one.

11. (a) 2,8,6
or
a correct target diagram

(b) $Mg(g) \rightarrow Mg^+(g) + e^-$
$Mg \rightarrow Mg^+ + e$
$Mg(g) \rightarrow Mg^+ + e^-$
$Mg \rightarrow Mg^+(g) + e$
or
$Mg(g) - e^- \rightarrow Mg^+(g)$
etc.

(c) Decreases
or
As you go from lithium to potassium (alkali metals) it (ionisation energy) decreases.
or
As you go from fluorine to bromine (halogens) it (ionisation energy) decreases.
or
As the atomic number in the group increases it decreases

12. (a) But-2-ene
or
2-butene

(b) (Molecules/compounds /hydrocarbons/alkenes) with same molecular/chemical formula but a different structural formula

(c) Correct structural formula for 3-methylpent-2-ene
or
2 ethyl but-1-ene
e.g.

or mirror images
or correct shortened structural formula e.g.
$CH_3CHC(CH_3)CH_2CH_3$

13. (a) Carboxyl

(b) (i) Condensation (polymerisation)

(ii)

or mirror images
Accept full or shortened structural formula or combination of both.

14. (a) (i) Carbon monoxide **or** CO/2CO

(ii) Covalent

(b) Distillation/distilling

(c) The sodium **or** chlorine **or** products can be recycled/reused
or
Chlorine can be used in the first step
or
Sodium can be used in final step

15. (a) 16

(b) $0.0032/3.2 \times 10^{-3}$ **or** correctly rounded answer

16. 1 mark: The student has demonstrated a limited understanding of the chemistry involved. The candidate has made some statement(s) which is/are relevant to the situation, showing that at least a little of the chemistry within the problem is understood.
2 marks: The student has demonstrated a reasonable understanding of the chemistry involved. The student makes some statement(s) which is/are relevant to the situation, showing that the problem is understood.
3 marks: The maximum available mark would be awarded to a student who has demonstrated a good understanding of the chemistry involved. The student shows a good comprehension of the chemistry of the situation and has provided a logically correct answer to the question posed. This type of response might include a statement of the principles involved, a relationship or an equation, and the application of these to respond to the problem. This does not mean the answer has to be what might be termed an "excellent" answer or a "complete" one.

Acknowledgements

Permission has been sought from all relevant copyright holders and Hodder Gibson is grateful for the use of the following:

The passage 'Ocean Dead Zones' taken from an article by Jessica Wurzbacher published on http://sailorsforthesea.org (Model Paper 1 Section 2 page 12);
An extract from the article 'Trouble in the periodic table', taken from page 15 of 'Education in Chemistry', January 2012 © Royal Society of Chemistry (Model Paper 1 Section 2 page 20);
Volatile Organic Compounds Label © B&Q (Model Paper 2 Section 2 page 9);
An extract from the article 'Bromine in Polar Regions' by Dr. Denis Pöhler, Institute of Environmental Physics, University of Heidelberg (Model Paper 2 Section 2 page 17);
An extract taken from the article 'Self Distilling Vodka' by Dave Ansell, published on http://thenakedscientists.com, January 2012 © The Naked Scientists (Model Paper 3 Section 2 page 15);
An extract from the article 'Clean coal technology comes a step closer' © Russell Deeks from BBC Focus: Science and Technology, April 2013 (2015 Section 2 page 12).

Hodder Gibson would like to thank SQA for use of any past exam questions that may have been used in model papers, whether amended or in original form.